NAVORD OP 3358 (VOLUME 1)

FIRST REVISION
CHANGE 2

U.S. Navy Submarine Torpedo Mark 16 Mod 8 Handbook

FOR OFFICIAL USE ONLY

PUBLISHED BY DIRECTION OF
COMMANDER, NAVAL ORDNANCE SYSTEMS COMMAND

THIS PUBLICATION SUPERSEDES NAVWEPS OP 3358 DATED 1 JULY 1965

15 DECEMBER 1970
CHANGED 1 MAY 1974

DEPARTMENT OF THE NAVY
NAVAL ORDNANCE SYSTEMS COMMAND
WASHINGTON, D. C. 20360

TORPEDO MARK 16 MOD 8
DESCRIPTION

NAVORD OP 3358 FIRST REVISION (VOLUME 1)
POSTAL CHANGE
PCH No. 1-1

1 February 1974

Page 1 of 1 Page

PUBLISHED BY DIRECTION OF
COMMANDER, NAVAL ORDNANCE SYSTEMS COMMAND

After complying with the instructions below, insert this POSTAL CHANGE between the front cover and title page.

The purpose of this POSTAL CHANGE is to provide information on charging line hoses.

NOTE: The portion of the text affected by the latest change is indicated by a vertical line in the outer margin of the page.

The following pages are provided to be inserted over the pages they correct. Do not remove any original pages from the publication.

1. Page ii
2. This POSTAL CHANGE will be cancelled by a future change to OP 3358 FIRST REVISION (VOLUME 1).

Requests for additional copies of POSTAL CHANGE PCH 1-1 to NAVORD OP 3358 FIRST REVISION (VOLUME 1) should be submitted to Commanding Officer, Naval Underwater Systems Center, Newport, Rhode Island 02840.

DEPARTMENT OF THE NAVY
NAVAL ORDNANCE SYSTEMS COMMAND
WASHINGTON, D. C. 20360

NAVORD OP 3358 (FIRST REVISION) VOLUME
TORPEDO MARK 16 MOD 8
DESCRIPTION

CHANGE
1 March 19

Page 1 of 1 Pa

PUBLISHED BY DIRECTION OF
COMMANDER, NAVAL ORDNANCE SYSTEMS COMMA

NAVORD OP 3358 (FIRST REVISION) VOLUME 1
is changed as follows:

TORPEDO MARK 16 MOD
DESCRIPTI

After the attached enclosures have been inserted, record this CHANGE on the change
record sheet and insert this change guide between the front cover and title page.

The purpose of this CHANGE is to correct errors and discrepancies, add safety information on charging lines and dollies, and to provide an alternate method for painting
exercise head.

> NOTE: On a changed page, the portion
> of the text affected by the latest change
> is indicated by a vertical line in the
> outer margin of the page. Changes to illus-
> trations are indicated by miniature pointing
> hands.

Remove and insert the following pages except where indicated.

1. Title and A
2. v thru viii
3. 1-5 and 1-6
4. 2-1 thru 2-6
5. 2-9 and 2-10

6. 2-29 thru 2-33/2-34
7. 2-39 and 2-40
8. 3-29 and 3-30
9. Index-5 and Index 6
10. Index-9 and Index-10

Requests for additional copies of CHANGE 1 to OP 3358 (FIRST REVISION) VOLUME 1
should be submitted to Commanding Officer, Naval Publications and Forms Center, 5801
Taber Avenue, Philadelphia 19120.

DEPARTMENT OF THE NAVY
NAVAL ORDNANCE SYSTEMS COMMAND
WASHINGTON, D. C. 20360

NAVORD OP 3358 (FIRST REVISION) VOLUME 1
CHANGE 2

1 MAY 1974

TORPEDO MARK 16 MOD 8
DESCRIPTION

PAGE 1 of 1 PAGE

PUBLISHED BY DIRECTION OF
COMMANDER, NAVAL ORDNANCE SYSTEMS COMMAND

After the attached enclosures have been inserted, record this CHANGE on the change record sheet and insert this change guide between the front cover and title page.

The purpose of this CHANGE is to delete all reference to the enabler or enabling system (ORDALT 10367). This CHANGE also cancels PCH 1-1 and PCH 1-2 which should be removed and destroyed.

> NOTE: On a changed page, the portion of the text affected by the latest change is indicated by a vertical line in the outer margin of the page. Changes to illustrations are indicated by miniature pointing hands.

Remove and insert the following pages:

1. Title and A
2. Foreword/Foreword-2
3. i thru iv
4. 1-1 thru 1-10
5. 2-19 thru 2-22
6. 2-49 and 2-50
7. 2-53 thru 2-56
8. 3-1 and 3-2
9. 3-7 thru 3-10
10. 3-23 and 3-24
11. Index-1 thru Index-4
12. Index-7 thru Index-10
13. Distribution List

Requests for additional copies of CHANGE 2 to OP 3358 (FIRST REVISION) VOLUME 1 should be submitted to Commanding Officer, Naval Publications and Forms Center, 5801 Tabor Avenue, Philadelphia, Pennsylvania 19120.

LIST OF EFFECTIVE PAGES

Insert latest changed pages; dispose of superseded pages in accordance with applicable regulations.

Page No.	# Change No.	Page No.	# Change No.
Title and A..................	2	2-30 and 2-31................	0
Foreword/Foreword-2..........	2	2-32.........................	1
Change Record/		2-33 and 2-34................	1
Change Record-2.........	0	2-35 thru 2-39..............	0
i thru iii..................	2	2-40........................	1
iv and v....................	0	2-41 thru 2-49..............	0
vi and vii..................	1	2-50........................	2
viii........................	0	2-51 and 2-52...............	0
1-1.........................	2	2-53 thru 2-56..............	2
1-2.........................	0	2-57 thru 2-62..............	0
1-3 thru 1-9................	2	3-1 and 3-2.................	2
1-10 thru 1-12..............	0	3-3 thru 3-6...............	0
2-1 and 2-2.................	1	3-7 thru 3-10..............	2
2-3.........................	0	3-11 thru 3-22..............	0
2-4.........................	1	3-23........................	2
2-5.........................	0	3-24 thru 3-29..............	0
2-6.........................	1	3-30........................	1
2-7 and 2-8.................	0	3-31 and 3-32...............	0
2-9.........................	1	Index-1 thru Index-3........	2
2-10 thru 2-19..............	0	Index-4.....................	0
2-20........................	2	Index-5.....................	1
2-21........................	0	Index-6.....................	0
2-22........................	2	Index-7 thru Index-9........	2
2-23 thru 2-28..............	0	Index-10....................	0
2-29........................	1	Distribution List...........	2

Zero in this column indicates an original page.

NAVORD OP 3358 (FIRST REVISION)

FOREWORD

NAVORD OP 3358 (FIRST REVISION) describes Torpedo Mk 16 Mod 8 and provides the instructions and data necessary to operate and maintain this equipment.

This publication consists of three volumes.

Volume 1 - Description
Chapter 1 - Introduction
Chapter 2 - Physical Description
Chapter 3 - Functional Description
Index

Volume 2 - Preparation for Use and Maintenance
Chapter 4 - Preparation for Torpedo Run
Chapter 5 - Troubleshooting and Repair
Chapter 6 - Postrun Procedures
Chapter 7 - Maintenance and Overhaul
Chapter 8 - Navol Handling
Chapter 9 - Equipment, Tools, and Material
Index

Volume 3 - Firing Craft Procedures
Chapter 10 - Loading and Off-Loading
Chapter 11 - Patrol Maintenance
Chapter 12 - Final Preparation
Chapter 13 - Accidental Activation Procedures
Chapter 14 - Navol Handling
Chapter 15 - Equipment, Tools, and Material
Index

This publication supports the basic torpedo with the following ORDALTS accomplished:

ORDALT	Description
3681	Modifies Gyro Adjusting Stand WA 2490
3711	Modifies Functional Test Air Muffler Assembly; LD 160704 Dwg 692844
3712	Deleted by CHANGE 2
3894	Modifies Spray Nozzles and Restrictions Test Stand Assembly; LD 160608 Dwg 768364
10195	Installs Cable Guide in the Control Mechanism Assembly to Prevent Damage to the Cable by the Propeller Shaft and Pallet Driving Gears
10203	Modifies Aft Propeller Nut to Permit Attachment of a New Propeller Guard
10236	Installs a Fuel Filter Assembly
10367	Remove the Enabler
10390	Depth Mechanism Test Stand

Ships, training activities, supply points, depots, Naval Shipyards, and Supervisors of Shipbuilding are requested to arrange for the maximum practicable use and evaluation of the standard ordnance technical manuals and to advise the Commanding Officer, Naval Underwater Systems Center, Newport, Rhode Island 02840 of all errors, omissions, discrepancies, and suggested improvements.

CHANGE RECORD

Change No.	Dated	Inserted By	Date Inserted

TABLE OF CONTENTS

LIST OF ILLUSTRATIONS

LIST OF TABLES

CHAPTER 1

INTRODUCTION

1-1. GENERAL DESCRIPTION

1-2. Torpedo Mk 16 Mod 8 (figure 1-1) is an electrically started, high-speed, long-range weapon. The combustion of Navol (hydrogen peroxide) and alcohol produces high-velocity steam, which drives a dual turbine engine. The turbines drive two counter-rotating propellers. The exhausted steam goes into the sea and condenses, resulting in a practically invisible wake. Prior to launch, the submarine fire control system presets the running pattern into the torpedo control mechanism; the control mechanism then steers the torpedo on this preset course.

1-3. The torpedo operates in either an exercise or a warshot condition. The exercise condition uses an exercise head for fleet training and torpedo proofing; the warshot condition uses a high-explosive warhead for destroying enemy surface ships.

1-4. TACTICAL EMPLOYMENT. Impulse pressure from the firing craft tube fires the torpedo. Three preset elements make up its running pattern: course, enabling run, and depth. Refer to figure 1-2.

1-5. Course. The torpedo runs a course that is any angle within 155° to the left (port) or right (starboard) of the submarine heading. (This can be extended to 157° if the course is set mechanically.) If for any reason the torpedo attains a launch angle greater than 165° during the initial 55 seconds of the run, an anti-circling-run (ACR) device shuts down the torpedo.

1-6. Running Depth. The firing vessel can launch the torpedo at any depth between 10 and 200 feet below the surface, though the optimum launch depth is between 10 and 100 feet. The running depth, which is preset, is limited to between 10 and 50 feet.

1-7. Deleted by CHANGE 2.

1-8. EXERCISE RUN. The exercise head contains 577 pounds of fresh-water ballast, giving the torpedo a negative buoyancy of 970 pounds. Toward the end of the run, air pressure forces the water ballast out, and the torpedo attains positive buoyancy, surfacing after 5.7 minutes. By adding a dye to the water ballast, the recovery vessel can more easily locate the torpedo at the end of the run; pingers may also be installed in the torpedo to help locate it, should it malfunction and sink. A depth and roll recorder produces a permanent record for evaluating torpedo stability and operation of its depth mechanism.

1-9. MAJOR COMPONENTS

1-10. The complete torpedo consists of three major sections: the warhead (or the exercise head), the energy section, and the afterbody and tail section.

1-11. COMPLETE TORPEDO. The warshot configuration of the torpedo is 246 inches long and weighs 3782 pounds; the exercise configuration is 244 inches long and weighs 3444 pounds. The torpedo is 21 inches in diameter. Refer to figure 1-3.

1-12. COMPONENTS. Figure 1-4 shows the location of major components, and table 1-1 briefly explains their functions.

1-13. TORPEDO CHARACTERISTICS

1-14. Table 1-2 lists the physical characteristics of the torpedo.

1-15. REFERENCE DOCUMENTS

1-16. Table 1-3 lists publications containing information concerning the torpedo.

1-17. CHECKLISTS

1-18. Torpedoes will be maintained and prepared in strict accordance with mandatory checklists contained throughout this OP. It is the individual activity's responsibility to reproduce these checklists, as they will become a permanent part of the torpedo record and will accompany the torpedo when it is transferred. The checklists will be retained for one overhaul or repeat-firing overhaul.

1-19. There are two methods for using the checklists: the reader/worker method and the worker method. The following paragraphs describe the two methods and list checklists to be used with each method.

1-20. READER/WORKER METHOD. Two people must use this method; their responsibilities are outlined as follows.

1-21. Reader's Responsibility. The reader should:

1. Insure that steps for work to be performed are read accurately.

2. Read aloud all warnings, cautions, and notes as they occur.

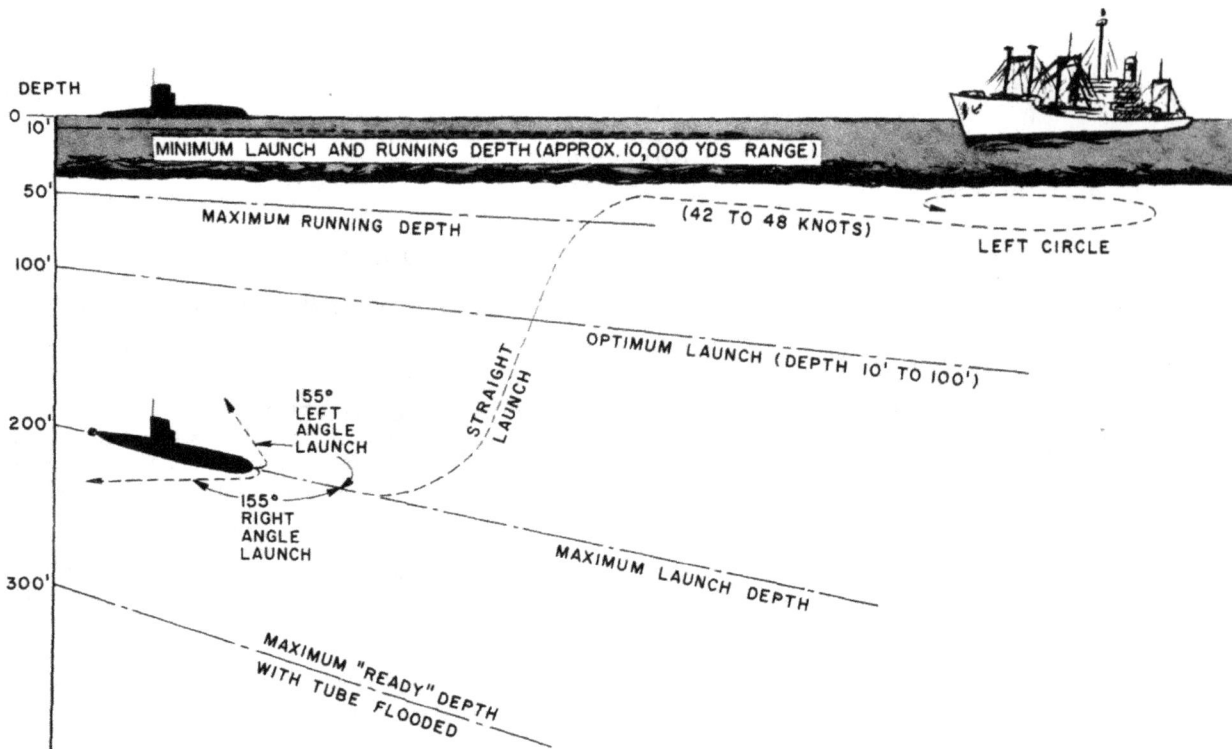

Figure 1-2. Running Pattern Characteristics.

3. Read aloud complete step, insuring that worker understands.

4. Observe worker's performance, when possible, as double check to insure step is being performed properly. (In some instances, it may be *more efficient to have second worker or observer* perform this function.)

5. Check off step when completed, after receiving a report, such as "check" from the worker.

6. Insure that all steps are performed in sequence, as listed on checklist.

7. After completing checklist, affix appropriate signatures as required for that particular evolution. (See paragraph 1-23.)

1-22. Worker's Responsibility. The worker should:

1. Listen to reading of step, insuring that all actions are understood prior to performing work.

2. Perform step as specified, using required tools.

3. Report completion of step, such as by saying "Check".

1-23. Checklists Requiring Reader/Worker Method. Table 1-4 lists the checklists that require use of the reader/worker method, indicating the signatures required upon completion. All other checklists may be performed by the worker method.

1-24. WORKER METHOD. This method requires frequent reference to the procedures by the worker to insure that no steps are omitted. In general, work must be performed in the sequence specified in the detailed procedures. It is recognized that, in the case of complex mechanical torpedoes, deviations in the sequence may be necessary. Such deviations may be authorized by the preparing activity.

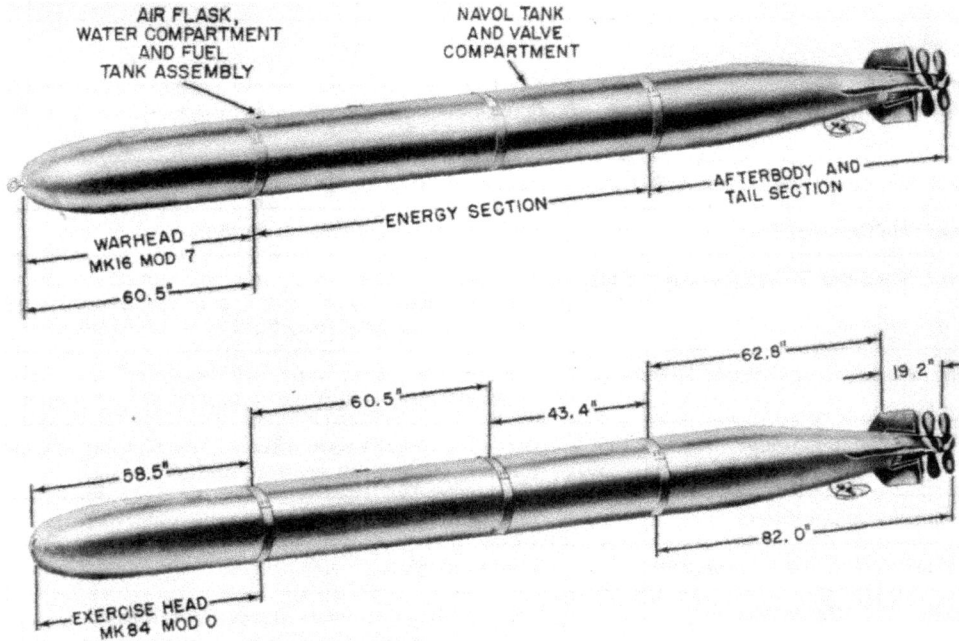

Figure 1-3. Torpedo Physical Characteristics

TABLE 1-1. MAJOR COMPONENTS

Fig. 1-4 Item No.	Component	Function
31	A-Cable	Conducts synchro signals, power-pack voltage, and firing impulse from fire control system to torpedo.
10	Air Flask, Water Compartment, and Fuel Tank Assembly	Stores air, water, and fuel for energy section.
49	Air Release Mechanism Mk 3 Mod 5	Releases air pressure to exercise head at end of run to expel water ballast.
41	Anti-Circling-Run (ACR) Device	Shuts down torpedo in event it circles back toward firing vessel.
39	Combustion System	Converts Navol, alcohol, and water into hot gases and steam to drive turbine.
34	Control Mechanism	Controls course and running depth of torpedo.
46	Depth and Roll (D & R) Recorder	Provides time record of running depth and torpedo roll changes during exercise run.
23	Electrical Cabling	Connects torpedo electronic components to A-cable connector; also connects Navol surveillance system to indicator panel.
32	Deleted by CHANGE 2	
4	Exploder Mk 9 Mod 7	Detonates warhead explosive charge.
21	Frame and Valves Assembly	Distributes air, Navol, water, and fuel.

TABLE 1-1. MAJOR COMPONENTS - Continued

Fig. 1-4 Item No.	Component	Function
26	Governor Assembly	Shuts down torpedo in event of overspeed (such as takes place out of water).
11	Main Air Pipe and Conduit Assembly	Houses main air pipe through center of torpedo.
24	Navol Monitoring Unit	Monitors gassing rate of Navol.
19	Navol Tank and Valve Compartment	Provides Navol storage; mounts propulsion system delivery valves, power pack, Navol monitoring unit, combustion system, and plug, manifold, and pipes assembly.
20	Plug, Manifold, and Pipes Assembly	Connects air flask, water compartment, and fuel tank to various valves in Navol tank and valve compartment.
40	Power Pack Assembly	Fires squib-activated valves in particular sequence for proper propulsion system operation.
29	Propeller Assemblies	Drives torpedo.
30	Rudders and Linkage Assembly	Guides torpedo.
3	Search Coil Mk 28 Mod 0	Provides for influence firing of warhead.
28	Tail Cone and Blades Assembly	Provides torpedo stabilization during run.
44	Torpedo Locating Device	Emits sonic signal to help locate sunken torpedo after exercise run.
38	Turbine and Gear Train	Provides rotary motion for turning propeller assemblies.

LEGEND FOR FIGURE 1-4

1 WARHEAD
2 EXPLOSIVE CHARGE
3 SEARCH COIL
4 EXPLODER
5 VENT VALVE
6 CLAMP RING SEGMENT
7 BLANKING NUT
8 HIGH-PRESSURE AIR PIPE ASSEMBLY
9 BLOW VALVE CONNECTING PIPE
10 AIR FLASK, WATER COMPARTMENT, AND FUEL TANK ASSEMBLY
11 MAIN AIR PIPE AND CONDUIT ASSEMBLY
12 AIR FLASK
13 WATER COMPARTMENT
14 FUEL TANK
15 WATER COMPARTMENT AIR INLET CONNECTION
16 FUEL TANK AIR INLET CONNECTION
17 FUEL OUTLET CONNECTION
18 WATER OUTLET CONNECTION
19 NAVOL TANK AND VALVE COMPARTMENT
20 PLUG, MANIFOLD, AND PIPES ASSEMBLY
21 FRAME AND VALVES ASSEMBLY
22 FUEL FILLING ACCESS OPENING
23 ELECTRICAL CABLING
24 NAVOL MONITORING UNIT
25 AFTERBODY AND TAIL SECTION
26 GOVERNOR ASSEMBLY
27 TEST CONNECTION
28 TAIL CONE AND BLADES ASSEMBLY
29 PROPELLER ASSEMBLIES
30 RUDDERS AND LINKAGE ASSEMBLY
31 A-CABLE
32 Deleted by CHANGE 2
33
34 CONTROL MECHANISM
35 HANDHOLE COVER
36 GYRO-SPIN AIR TUBE
37 EXHAUST MANIFOLD ASSEMBLY
38 TURBINE AND GEAR TRAIN
39 COMBUSTION SYSTEM
40 POWER PACK
41 ANTI-CIRCLING-RUN (ACR) DEVICE
42 NAVOL TANK
43 EXERCISE HEAD
44 TORPEDO LOCATING DEVICE
45 ACCESS COVER
46 DEPTH AND ROLL RECORDER
47 VENT VALVE
48 DISCHARGE VALVES
49 AIR-RELEASE MECHANISM

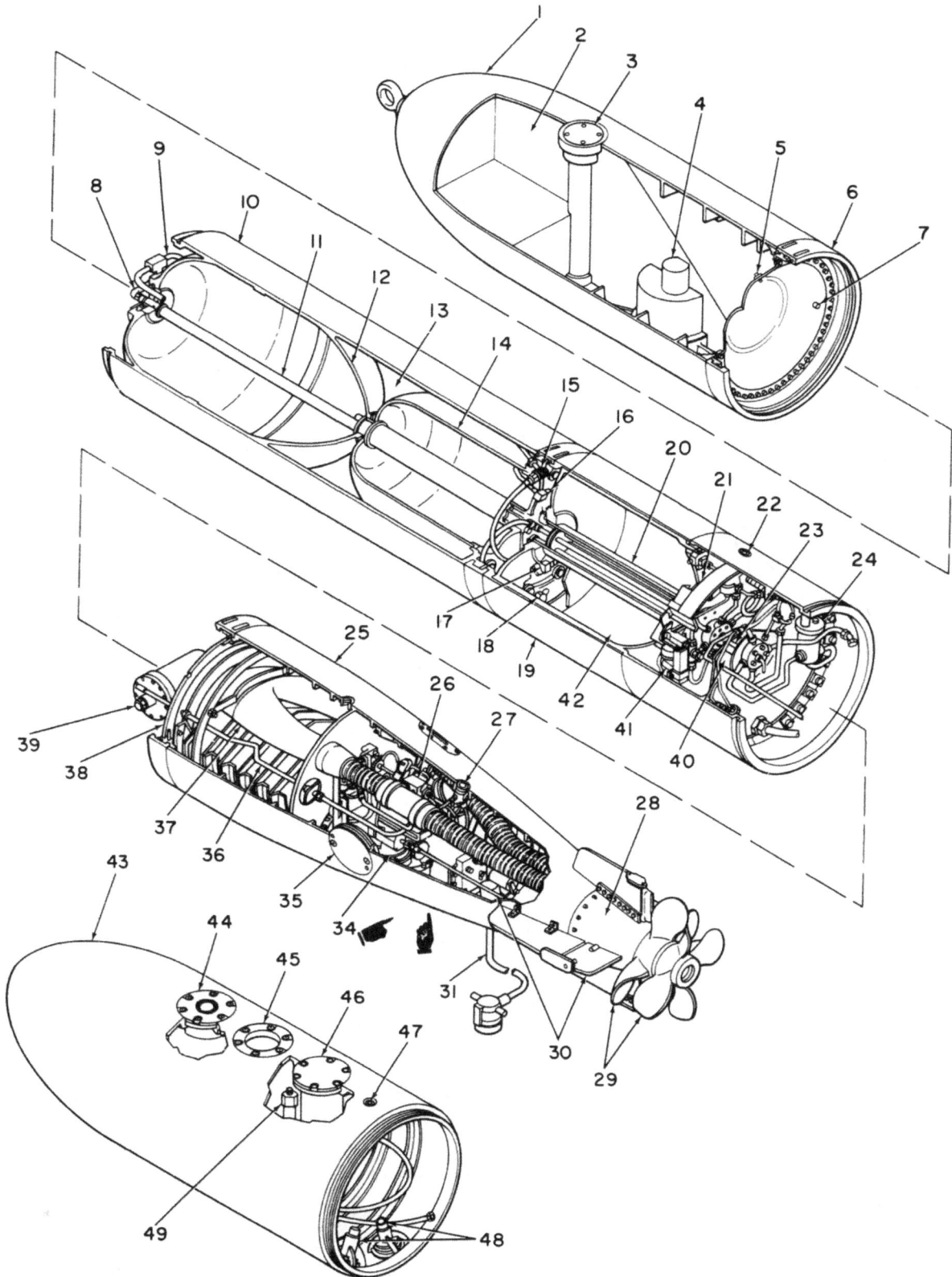

Figure 1-4. Torpedo Major Components.

TABLE 1-2.   TORPEDO CHARACTERISTICS

Dimensions	
Overall	
With warhead	20 ft 6. 5 in.
With exercise head	20 ft 4. 5 in.
Approximate Weight	
Warshot	3782 lb
Exercise shot	3444 lb
Exercise shot buoyancy (empty)	266 lb
Modified fully ready torpedo (headless)	2516 lb
Torpedo displacement (in sea water)	2471 lb
Warhead (fully ready)	1267 lb
Shell assembly (with search coil, joint ring, and bulkhead installed)	483 lb
Exploder	34 lb
Arming device	3 lb
Booster	1 lb
Explosive Charge HBX-3	747 lb
Displacement (in sea water)	666 lb
Exercise head (fully ready)	928 lb
Fresh water ballast	577 lb
Without fresh water ballast	351 lb
Depth and roll recorder	16 lb
Air release mechanism	6 lb
Torpedo locating device (pinger assembled)	6 lb
Center of Gravity (from tail end) with propeller guard installed.	
Warshot	144. 3 in.
Exercise shot	136. 5 in.
Modified Fully Ready (Headless)	107. 5 in.
Dynamic Characteristics	
Engine horsepower (dynamometer)	350
Turbine wheel speed	12, 870 rpm

TABLE 1-2. TORPEDO CHARACTERISTICS - Continued

Dynamic Characteristics - Continued	
Gear ratio (turbine to propeller)	9:1
Propeller speed	1430 rpm
Nozzle working temperature	1200° F
Overall range (10-50 feet running depth)	
Warshot	10,000 yd
Exercise shot	10,000 yd
Speed (15 feet running depth)	
0 to 1000 yards of run	43.6 kn
1000 yards to end of run	46.2 kn
Gyro angle (deflection corrected at 5000 yards, with 0° setting)	6 mils R±12
Tank Capacities	
Air flask (2800 psi)	68 lb, 4.81 cu ft
Water compartment (distilled or battery water)	218 lb, 3.4 cu ft
Fuel tank (92.5% alcohol)	103 lb, 2.02 cu ft
Navol tank (70% hydrogen peroxide)	292 lb, 3.58 cu ft
Engine casing (oil)	9 lb, 0.17 cu ft
Fluid Outage	
Navol	20 lb, 14.5 pints
Unit Pressures	
Air flask (working)	2800 psi
Water compartment (working)	635 psi
Fuel tank (working)	635 psi
Navol tank (working)	635 psi
Reducing valve (working)	635 psi
Nozzle (working)	485 psi
Horizontal and vertical steering engines (working)	635 psi
Deleted by CHANGE 2	
Gyro constant spin (working)	125 psi
Air release mechanism (opening)	1450 psi

TABLE 1-2. TORPEDO CHARACTERISTICS - Continued

Test Pressures	
Afterbody	16 psi
Exercise head	15 psi
Warhead	5 psi
Exploder cavity	5 psi
Fluid Flow Rates (50 ±0.5 pounds per volume)	
Water	141 ±2 sec, 45 ±0.5 psi
Alcohol	241 ±2 sec, 90 ±0.5 psi
Navol	77 ±1 sec, 35 ±0.5 psi
Maximum Righting Moments	
Warshot	2756 in.-lb
Exercise shot	1575 in.-lb
Warhead (with search coil, cable, and joint ring attached - otherwise empty)	1638 in.-lb
Warhead (without exploder and arming device)	1911 in.-lb
Control Settings	
Rudder throws	
Horizontal-neutral position	1.5 graduations down
-up position	1.0 graduations up
-down position	4.0 graduations down
Vertical -upper (vernier scale in left or right position)	19.0 ±1
-lower (vernier scale in left or right position)	13.0 ±1
Gyro angle launch (left or right, maximum)	155.0° (electrically) 157.0° (mechanically)
Deleted by CHANGE 2	
Left circling (radius)	315 ±50 yd
Depth (running, minimum and maximum)	10-50 ft
Designation of Components by Mk and Mod	
Warhead	16-7
Search coil	28-0

TABLE 1-2. TORPEDO CHARACTERISTICS - Continued

Designation of Components by Mk and Mod - Continued	
Exploder	9-7
Arming device	3-2
Primer (instant)	113-0
Primer (1/3-second delay)	149-0
Booster	5-0
Hydrostatic switch	21-2
Hydrostatic switch (floor)	72-0
Inertia switch	3-3
Exercise Head	84-0
Depth and roll recorder (D & R)	2-1
Air release mechanism	3-5
Gyro	12-3
Igniter	6-4
Servomotor (control mechanism)	7-3
Deleted by CHANGE 2	
Synchro transmitter (control mechanism)	22-2
Synchro transmitter (enabler)	22-1
Torpedo control A-cable	1-12 or 1-13

TABLE 1-3. REFERENCE DOCUMENTS

Publication No.	Title
OP 627	U.S. Navy Torpedo Gyroscopes, Non-Tumble Type
OP 1105 (Vol 6)	Preservation and Preservation Maintenance of Ordnance Equipment in Shore Storage; Underwater Ordnance
OP 1303	U.S. Navy Synchros; Description and Operation
OP 1711	Depth and Roll Recorder Mark 2 Mod 0
OP 2139	Indicator Panel Mark 25 Mod 1

TABLE 1-3. REFERENCE DOCUMENTS - Continued

Publication No.	Title
OP 2473	Indicator Panel Mark 25 Mods 2 and 3; Description, Operation, and Maintenance
OP 2639	Torpedo Control Cables and Associated Equipment; Description, Installation, and Maintenance
OP 2744	Indicator Panel Mark 25 Mod 4; Description, Operation, and Maintenance
OP 3097	Indicator Panel Mark 21 Mod 2; Description, Operation, and Maintenance
OP 3347	U.S. Navy Ordnance Safety Precautions
OP 3370	Torpedo Exploder Mark 9 Mod 7; Description, Operation, and Maintenance
OP 3525	Test Set Mark 281 Mod 1; Description, Operation, and Maintenance
OD 3000	Lubrication of Ordnance Equipment
OD 6279	Torpedoes Marks 13, 14, 15, 16, 17, 18, and 23 Types; Pressure, Temperature, and Humidity Effect on Torpedo Running Depths
OD 6470	Instructions for Adjustment and Maintenance of Leakage Test Stand WA 16081; for Torpedoes Mark 14 and Mark 16
OD 9019	Magnetic Balancing of Warhead Mark 16 Mod 7
OD 9092	Fire Control Switch Box Mark 8 Mod 0; Functions and Instructions
OD 9093	Synchro Zeroing Device
OD 10235	Immersion Gear Casing and Pendulum, Alignment Gages; Description and Test Procedure
OD 10765	Torpedo Workshop Equipment; Test Set Mark 259 Mod 1; LD 620085
OD 12404	Torpedo Workshop Equipment; Universal Dolly and Adapters
OD 13088	Torpedo Workshop Equipment; Spray Nozzles and Restrictions Test Stand, BUWEPS LD 160608 (U)
OD 13089	Torpedo Workshop Equipment; General Use Test Panels LD 140423 and LD 495804 (U)
OD 13092	Torpedo Workshop Equipment; Torpedo Storage Racks (U)
OD 13103	Synchro Zeroing Device Mark 321 Mod 0
OD 13113	Test Set Mark 255 Mod 0; Description, Operation, and Maintenance
OD 15165	Torpedo Mark 16 Mod 8; Air Flask, Water Compartment, and Fuel Flask Assembly, LD 162338; Hydrostatic and Pneumatic Tests
OD 15176	Initial Application and Repair of Preservatives for Metallic Surfaces
OD 15182 (FIRST REV)	Emergency Hydrogen Peroxide (Navol) Stabilizing Process for Torpedo Mark 16

TABLE 1-3. REFERENCE DOCUMENTS - Continued

Publication No.	Title
OD 15196	Use of Depth and Roll Recorder Mark 2 Mods 0 and 1 to Evaluate Torpedo Mark 16 Mod 8 Run Performance (U)
OD 16086	U. S. Naval Underwater Weapons; Operational Characteristics and Tactical Data
BUWEPS INST. 5101.2	Hazards of Electromagnetic Radiation to Ordnance (HERO): Policy for Conduct of Program to Alleviate
BUWEPS INST. 8510.7B	Torpedo Firing Reporting System (Form 8510/2A)
BUWEPS INST. 8510.8A	Report of Unsatisfactory or Defective Torpedoes or Equipment (RUDTORPE), NAVWEPS Form 8510/3 (10-61) (Report Symbol BUWEPS 8510-3)
BUWEPS INST. 8510.9	Torpedo Overhaul (or Maintenance) Schedules
BUWEPS INST. 8510.25	Bureau of Naval Weapons (BUWEPS) managed Torpedo, ASROC, and SUBROC 4T Cognizance Material; requisitioning procedures for
NAVORD INST. 8510.8	Exercise Heads for Steam and Acoustic Type Torpedoes; Allowances of
NAVORD INST. 8510.9	Torpedo Exploder History Record, NAVORD Form 8510/4 (Report Symbol NAVORD 8510-4); Establishment of
NAVORD INST. 8510.10	Torpedo Igniter Mark 6 Mod 4; Maintenance, Disposition, and Replacement of
NAVORD INST. 8510.12	Allowance per Weapon of Torpedo and ASROC Components
NAVORD REPORT NO. 543	Navol Loading Depot
NAVORD REPORT NO. 5216	Introduction to Hydrogen Peroxide
NAVSHIPS 0975-000-0010	SSN 637 Class Operations Manual
NAVSHIPS 0975-000-1010	SSN 594 Class Operations Manual
OPNAV INST. 8023.7	Rules and Regulations for Military Explosives and Hazardous Munitions
SPCC INST. 4440.83	Policy and Procedures for Handling Cognizance Symbol 2A Repairable Ordnance Material

TABLE 1-4. CHECKLISTS REQUIRING READER/WORKER METHOD

Checklist No.	Title	Signature Required
4-8	Afterbody and Tail Section System Test	Shop Supervisor or OIC
4-9	Final Assembly of Energy Section	Shop Supervisor or OIC

TABLE 1-4. CHECKLISTS REQUIRING READER/WORKER METHOD - Continued

Checklist No.	Title	Signature Required
4-10	Assembly of Afterbody and Tail Section to Energy Section	Shop Supervisor or OIC
4-11	Assembly of Exercise Head to Energy Section	Witnessing Officer from firing craft or representative
4-12	Assembly of Warhead to Energy Section	Witnessing Officer from firing craft or representative
	Firing Craft Procedures*	Witnessing Officer aboard firing craft

*Firing craft procedure checklists are contained in torpedo checklist OD 44979.

CHAPTER 2

PHYSICAL DESCRIPTION

2-1. WARHEAD MK 16 MOD 7

2-2. Warhead Mk 16 Mod 7 (figure 2-1) is issued for use with Torpedo Mk 16 Mod 8. In its issued condition, the warhead consists of a shell assembly containing a main explosive charge (HBX-3), lead ballast, Search Coil Mk 28 Mod 0, an eyebolt, and a cavity for receiving Exploder Mk 9 Mod 7 and Booster Mk 5 Mod 0. The exploder uses Arming Device Mk 3 Mod 2. The exploder, arming device, and booster are issued in separate containers and are installed in the warhead when the torpedo is made fully ready. (If the exploder is temporarily stowed in the warhead, the booster and arming device are installed when the torpedo is made fully ready.) The search coil is permanently installed and magnetically balanced in relation to the mass of the warhead.

2-3. SHELL ASSEMBLY. The warhead shell is made of a single piece of hydrodynamically shaped

sheet phosphor bronze, to which is added a nosepiece, an eyebolt, lead ballast, a search coil housing, an exploder cavity, 3 strengthening rings, a joint ring, and a bulkhead assembly. The bronze nosepiece is welded to the shell. The eyebolt, threaded into the nosepiece, permits convenient handling of the warhead and is removed to allow the torpedo to fit into the tube. The lead ballast, in the lower forward portion of the shell, provides roll and longitudinal stability. The main explosive charge is poured on a 60° angle, with the greater percentage of the explosive charge below the centerline, thus contributing to roll and longitudinal stability.

2-4. The search coil housing consists of a top flange and a bottom flange, to which is welded a central tube. Both flanges are welded to the internal surface of the warhead shell so that the housing is in a vertical position, slightly forward of the warhead center. A cover plate and gasket

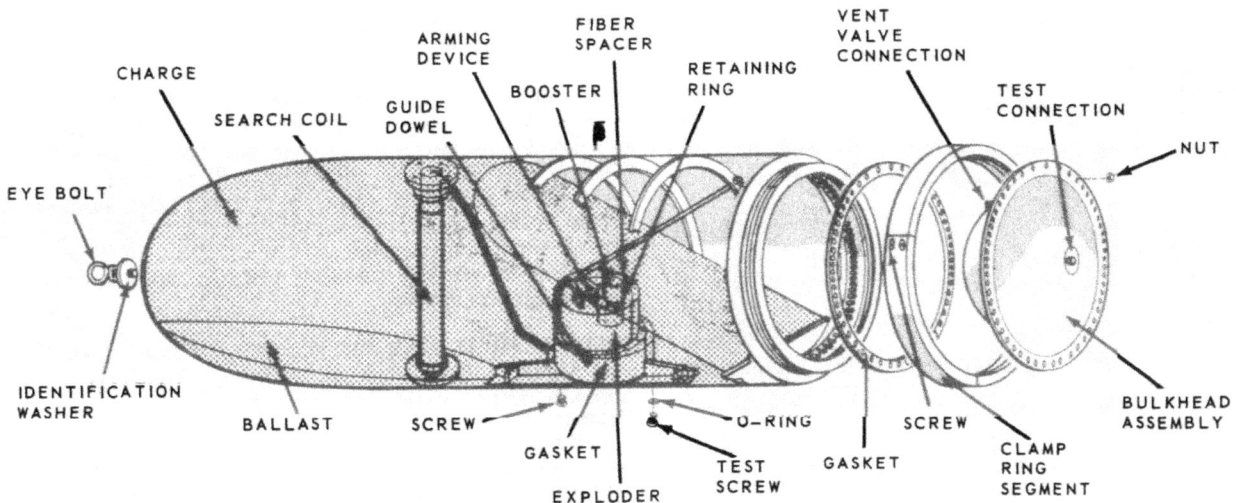

Figure 2-1. Warhead Mk 16 Mod 7.

seals each of the flange openings. A tube from the top flange to the exploder cavity completes the search coil housing. The search coil is installed in the central tube, interconnected to the exploder by the search coil cable, which runs through the tube between the top flange and exploder cavity.

2-5. The exploder cavity consists of a bronze circular flange and copper casing welded assembly installed by having the flange welded to the internal surface of the warhead shell so that the cavity is on the bottom centerline and slightly aft of the warhead center. There is a channel in the base of the flange and in a portion of the warhead shell. As the torpedo moves through the water, this channel directs the water to the impeller of the exploder (figure 2-2), thus providing the driving power for certain exploder components. When the exploder is not installed, a cover plate and gasket is secured in the exploder cavity flange to protect and seal the cavity. A test screw and connection (tapped hole) in the flange channel allows vacuum- and pressure-testing of the exploder cavity and search coil housing. Tapped holes in the circular rim of the flange secure the exploder in place. A copper cylinder is brazed to the upper portion of the exploder cavity casing to completely house the booster and partially house the arming device. Dowel pins in the exploder cavity flange insure correct installation of the exploder and special gasket.

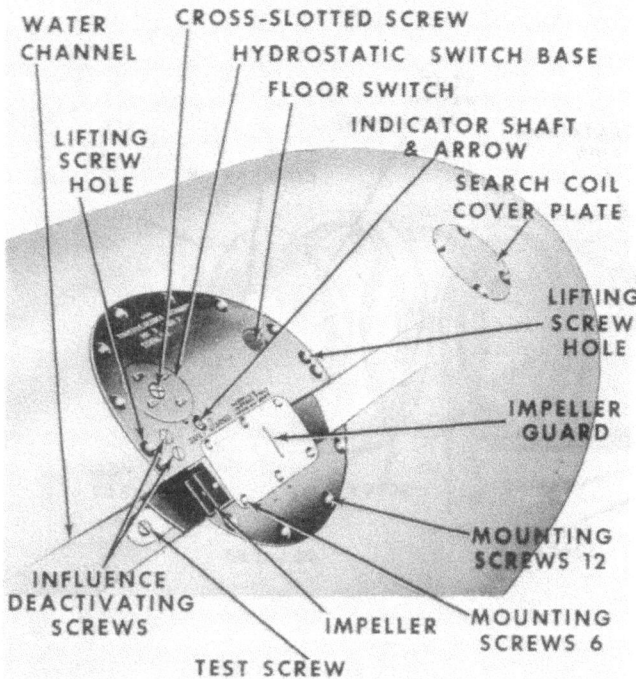

Figure 2-2. Exploder Mk 9 Mod 7 Installed in Warhead Mk 16 Mod 7.

2-6. Three bronze strengthening rings are welded to the internal surface of the warhead shell. These rings provide extra structural strength to the shell.

2-7. A segment of a circular piece of sheet bronze is welded to the lower portion of the warhead shell, just aft of the strengthening ring, at an angle of 60° from the horizontal centerline of the warhead. This piece of sheet bronze forms a dam to retain the angular load of the main explosive charge and to prevent the thin edge of the angular load from breaking.

2-8. The wedge-type steel joint ring at the after-end and perimeter of the warhead shell couples the warhead and the energy section; this provides additional structural strength for the shell, as well as a flange for attaching the bulkhead assembly. A protective ring is clamped to the joint ring by using three clamp-ring segments and six tangentially arranged screws. This protecting ring prevents damage to bearing and O-ring surfaces of the joint ring when the warhead is separated from the energy section.

2-9. The bulkhead assembly is a dome-shaped, circular piece of steel containing a vent valve and a test connection, mounted 90° from one another, and 52 equally spaced holes around its perimeter. The vent valve, an O-ring type, prevents external pressure from entering the warhead interior and allows excessive pressure within the warhead to escape. The test connection permits vacuum testing the water tightness of the warhead. A blanking cap, installed on the outer side of the test connection, seals the connection when not in use. The bulkhead assembly (and a gasket) seal the afterend of the warhead when they are secured to the joint ring with 52 equally spaced screws.

2-10. MAIN EXPLOSIVE CHARGE. The main explosive charge consists of approximately 747 pounds of HBX-3 and occupies almost the entire internal area of the warhead. The HBX-3 is cast into the warhead in a molten state so that the after surface of the explosive charge forms an angle of 60° from the horizontal centerline of the warhead. Impacts produced by normal handling will not detonate the main explosive charge.

2-11. SEARCH COIL MK 28 MOD 0. The search coil (figure 2-3) consists of two vertically stacked, opposite-wound coils that must be magnetically balanced when a uniform magnetic field is applied to the mass of a fully assembled warhead. Magnetic balancing is carried out immediately after the warhead is manufactured and shall not be disturbed during any phase of Fleet maintenance of the warhead.

2-12. BOOSTER MK 5 MOD 0. The booster is a charge of tetryl pellets housed in a cylindrical copper container. A fiber-bound pad and the booster are installed in the upper (smaller) cylindrical area of the exploder cavity before installing the exploder.

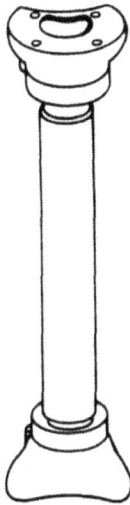

Figure 2-3. Search Coil Mk 28 Mod 0.

2-13. ARMING DEVICE MK 3 MOD 2. The arming device consists of an arming and safe rotary cam switch, an instantaneous electric primer, a 1/3-second delay electric primer, two powder trains (each having two portions), and a tetryl pellet charge -- all housed in a cylindrical container. When the rotary cam switch is in its armed position, both portions of each powder train are in alinement with each other, and each powder train is in alinement with one of the two electric primers. The tetryl pellet charge is always in alinement above the upper portions of both powder trains. Upon installation of the exploder in the warhead, the arming device is inserted in the well at the top of the exploder in alinement with Booster Mk 5 Mod 0.

2-14. EXPLODER MK 9 MOD 7. The exploder (figure 2-4) is nearly cylindrical in shape, approximately 8 inches in diameter, 8 inches in height, and weighs approximately 34 pounds. Its housing is made of bronze, curved to conform to the warhead contour. In addition to the arming device, it contains an impeller-driven alternator, an impedance-change transformer circuit, a power transformer circuit, an amplifier circuit, two firing thyratrons, Hydrostatic Switch Mk 21 Mod 2, Floor Switch Mk 72 Mod 0, and Inertia Firing Switch Mk 3 Mod 3.

2-15. EXERCISE HEAD MK 84 MOD 0

2-16. Exercise Head Mk 84 Mod 0 (figure 2-5) is used in torpedo proofing and Fleet exercise firing of Torpedo Mk 16 Mod 8. In its issued, or modified fully ready, condition, the main components of the exercise head consist of a shell assembly, lead ballast, two discharge valves, a vent valve, and Air Release Mechanism Mk 3 Mod 5. After the exercise head is attached to the torpedo and is prepared in its fully ready condition, two accessories - Depth and Roll Recorder Mk 2 Mod 1 and Torpedo Locating Device if used, are installed in designated shell

Figure 2-4. Exploder Mk 9 Mod 7.

flanges; water ballast completely fills the remaining internal shell space.

2-17. SHELL ASSEMBLY. The exercise head shell assembly consists of a hydrodynamically shaped, sheet-steel, cylindrical hull, curved at the forward end, to which is welded a steel nosepiece, several steel strengthening rings, several steel flanges and bosses, and a steel joint ring. In addition, approximately 140 pounds of lead ballast is angularly poured

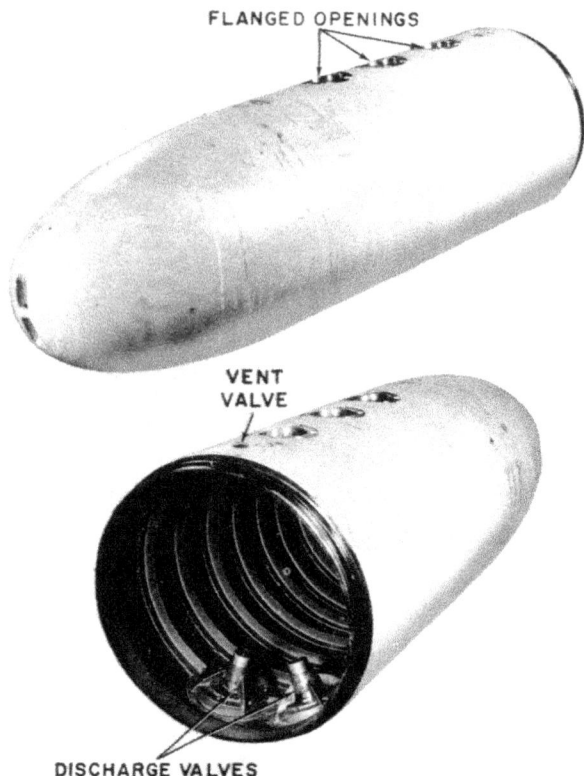

FLANGED OPENINGS

VENT
VALVE

DISCHARGE VALVES

Figure 2-5. Exercise Head Mk 84 Mod 0.

in the forward-bottom end of the shell. All un-machined external surfaces of the shell assembly are primed with a coat of zinc chromate. The shell is then painted with a coat of white enamel, a coat of luminescent orange-colored material, then a coat of lacquer. As an alternate method, shell may be painted with a coat of orange enamel over the primer. All unmachined internal surfaces of the shell assembly are coated with an air-drying phenolic resinoid material.

2-18. There are three flanged openings of the same size and in alinement on the top centerline of the shell. Two of the openings (the fore and aft ones) accommodate the accessories. The center opening provides access to internal parts of the exercise head when the head is attached to the energy section. Each flange of the three openings contains six studs and an O-ring groove. When an opening is not in use, it is sealed with an O-ring and blanking cover, held in place by a cover retainer and six nuts. When an accessory is installed, it is sealed to the flange by an O-ring and is held in place by its special retainer and six nuts.

2-19. A steel ring is welded to a shell strengthening ring just aft of the center flanged opening. The air release mechanism, upon installation, snaps onto this ring.

2-20. There are two slightly larger flanged openings similar to each other, radially side by side, located at the bottom-aft end of the shell. The discharge valves are installed in these openings. Each

flange forms the body of a discharge valve. Working components of each discharge valve are held in place by a retaining spring plate secured to six flange studs with six nuts.

2-21. A small threaded boss is on the top centerline near the aft end of the shell. The vent valve (figures 2-5 and 2-7) screws into this. Another slightly larger threaded boss on the bottom centerline, just forward of the discharge valve flanges, contains a removable plug for draining.

2-22. Thirteen strengthening rings, spaced along the length of the shell interior, provide additional structural strength necessary to prevent collapse or other damage when the shell is subjected to deep sea pressures. The joint ring provides structural strength, in addition to attaching the exercise head to the energy section. This joint ring has an O-ring groove that fits into the joint ring of the energy section. The joint rings of the two sections are held in place by three clamp-ring segments, similar for all wedge-type joints of the torpedo. A protecting ring prevents damage to the O-ring groove and bearing surfaces of the joint ring. The protecting ring is installed in the same manner as a mating joint ring of a wedge-type joint.

2-23. An eye in the nosepiece is used for handling the unattached exercise head or the floating torpedo during recovery after a run. The eye, however, is not strong enough to lift a complete torpedo.

2-24. DISCHARGE VALVES. The two identical discharge valves (figure 2-6) are essentially spring-loaded check valves that remain closed to prevent sea water from entering the exercise head throughout a normal torpedo exercise run, but open to expel the water ballast when the air-release mechanism releases air-flask pressure into the exercise head at the end of a torpedo exercise run.

2-25. VENT VALVE. The vent valve (figure 2-7) is a small, stainless steel valve, threaded at one end. The valve, grooved at the other end to accept an external O-ring, has a vent passage from the threaded end to the O-ring groove.

2-26. AIR RELEASE MECHANISM MK 3 MOD 5.
The self-triggered air-release mechanism (figure 2-8) consists of a valve body containing an air strainer and a spring-loaded valve held in place in a central hole of the body by a bushing, valve seat, and retaining screw installed in the bottom of the body and by a spring, spring support, adjusting nut, and lockscrew installed in the top of the body. The spring load is adjusted by the adjusting nut which is threaded into the body and secured in place by the lockscrew. The air strainer is held in place by a holder installed in the bottom of the body and an air-inlet nipple and gasket installed in the holder. An O-ring covers air-outlet ports in the central perimeter of the body. When the air-release mechanism is installed in the exercise head, it is attached to a ring (paragraph 2-19), by a snaphook secured to the upper portion of the body, and the air-inlet nipple is connected to the outlet tube of

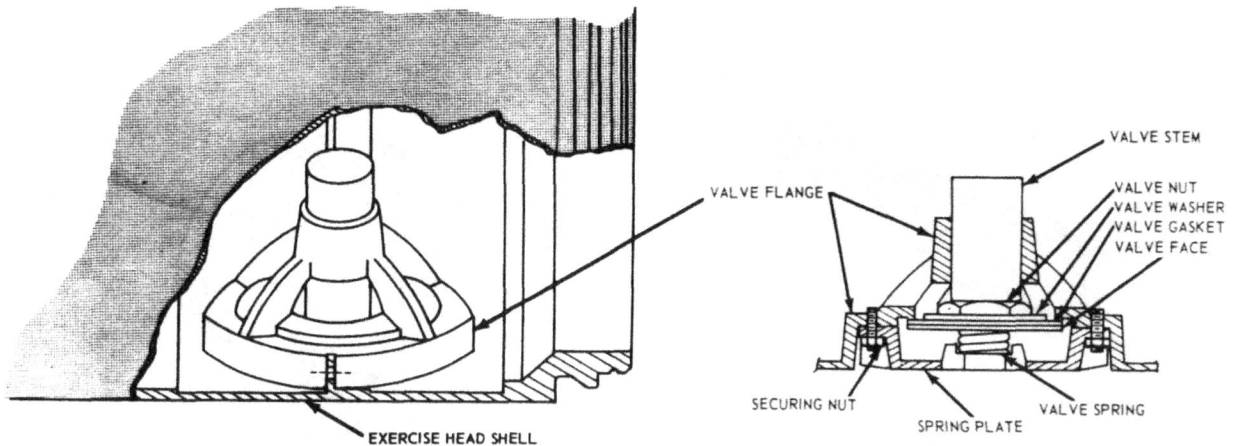

Figure 2-6. Discharge Valve.

Figure 2-7. Vent Valve.

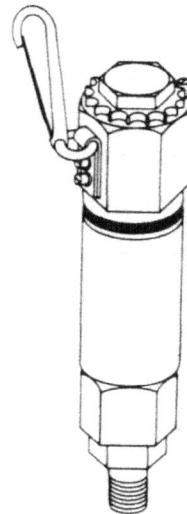

Figure 2-8. Air Release Mechanism.

the blow valve of the energy section by a high-pressure pipe assembly. The air-release mechanism has a brass body, a porous bronze filter, a nylon valve seat, and a metal valve and spring arrangement.

2-27. DEPTH AND ROLL RECORDER MK 2 MOD 1. The depth and roll (D&R) recorder (figure 2-9) consists of a sheet brass case containing a brass framework on which is mounted a sea-water depth-sensing mechanism, a roll pendulum, a recording mechanism, a gear train, a governor, and a spring-operated motor. The framework consists of a top circular piece, three thinner circular plates, and

a bottom support ring. Suitable supporting posts are secured to the ring perimeters to form four compartments. The uppermost compartment contains the depth-sensing mechanism, consisting of a bellows and linkage. The next lower (second) compartment contains the roll pendulum and its linkage, plus the recording mechanism. The recording mechanism uses moving carbon paper, scratched by a stylus of the depth and roll linkage. The third compartment contains a gear train and a governor to control the speed of the moving carbon paper. The bottom compartment contains a spring-operated motor, which has a flat coiled spring and two drums to drive the gear train.

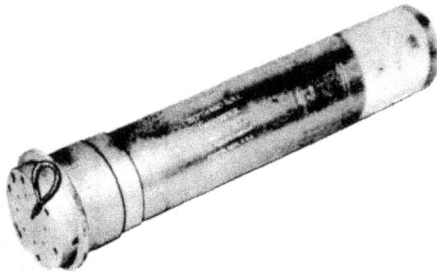

Figure 2-9. Depth and Roll Recorder.

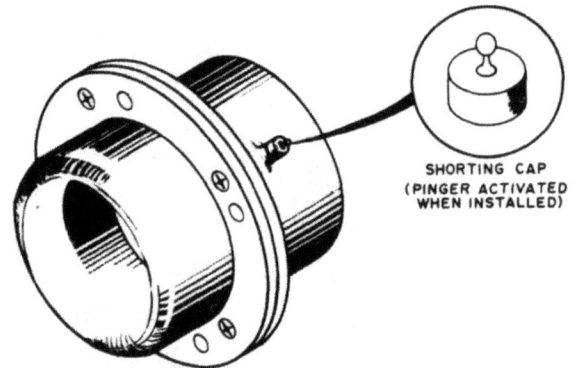

SHORTING CAP
(PINGER ACTIVATED
WHEN INSTALLED)

Figure 2-10. Torpedo Locating Device.

2-28. TORPEDO LOCATING DEVICE. Pinger 2815837 (figure 2-10) is a self-contained, battery-operated transmitting transducer, which emits ultrasonic acoustic energy (30-45 kHz) when activated. A shorting cap (male connector) is manually installed, prior to an exercise run, to activate the pinger and is removed for deactivation. The receiver system used to detect the ultrasonic pinger signals consists of a fully transistorized portable receiver in a watertight case which converts the ultrasonic energy to an audio frequency of approximately 2 kHz. An earphone is worn by a scuba diver when the receiver is hand carried, and an external hydraphone and headset is used for over-the-side use from a retrieving boat. The beamwidth of the receiving transducer is such that the audible signal is proportional to the bearing and range of the pinger. The pinger will operate for a period of 10 days and is installed in the exercise head and afterbody, when used.

2-29. ENERGY SECTION

2-30. The energy section is made up of two units: (1) air flask, water compartment, and fuel tank assembly, and (2) Navol tank and valve compartment. The units are interconnected by a wedge-type joint (the same as that used to interconnect torpedo main sections), a telescoping main-air pipe, and flexible pipes.

2-31. AIR FLASK, WATER COMPARTMENT, AND FUEL TANK ASSEMBLY. The air flask, water compartment, and fuel tank assembly (figure 2-11) stores high-pressure (2800 ±50 psi) air, water, and fuel (alcohol), as its name implies. It is composed of the following major components:

1. Flask Shell Assembly.

2. Main Air Pipe and Conduit Assembly.

3. Bulkhead Assembly.

4. Blow Valve Assembly.

5. Water Compartment Air-Inlet Connection.

6. Water-Outlet Connection.

7. Relief Valve Assembly.

8. Guide Studs.

9. Fuel Tank Assembly.

10. Lock-Ring Segments and Brace.

2-32. Flask Shell Assembly. The flask shell assembly (figure 2-11) consists of the completely assembled air flask, a portion of the water compartment, and a flanged bulkhead to receive the fuel tank assembly. The air flask is formed by the forward portion of the outer shell, two dome-shaped bulkheads, a bulkhead assembly, and the main pipe and conduit assembly. The aft portion of the outer shell, the aft face of the after dome-shaped bulkhead of the air flask, and the flanged bulkhead for installation of the fuel tank form a portion of the water compartment. When the fuel tank is installed, the water compartment area is complete. A small threaded hole, centrally located on the top centerline of the flask shell, vents air and drains water during hydrostatic tests of the air flask. Another threaded hole, located on the top centerline and just aft of the main guide stud, is used to fill and drain the water compartment. When these holes are not in use, they are sealed with gaskets and plugs.

2-33. Main Air Pipe and Conduit Assembly. The main air pipe and conduit assembly (figure 2-12) passes through the center and along the entire length of the air flask, water compartment, and fuel tank assembly. The conduit is the outer portion of the assembly and consists of a tube, to which are welded three plugs equipped with O-rings and backup rings. The forward plug inserts into another plug to seal the center hole in the forward bulkhead of the air flask. The center plug seals the after bulkhead of the air flask. The after

1. BLOW VALVE (STOP PLUG INSTALLED)
2. BLOW VALVE INLET NIPPLE
3. PIPE ASSEMBLY ATTACHED TO PLUG ASSEMBLY
4. BLOW VALVE OUTLET NIPPLE
5. BACK-UP RING
6. O-RING
7. AIR FLASK VENT HOLE
8. AIR FLASK SHELL ASSEMBLY
9. O-RING
10. BACK-UP RING
11. GUIDE STUD HOLES
12. WATER FILLING HOLE
13. BLANKED OFF PIPE
14. WATER COMPARTMENT AIR INLET CONNECTION
15. FUEL TANK AIR INLET CONNECTION
16. FLAT WASHER
17. SECURING SCREW
18. O-RING
19. MAIN AIR PIPE
20. BRACE
21. FUEL OUTLET CONNECTION
22. WATER OUTLET CONNECTION
23. RETAINING-RING SEGMENT
24. O-RING
25. FUEL TANK
26. RETAINING RING
27. WATER COMPARTMENT
28. MAIN AIR PIPE AND CONDUIT ASSEMBLY
29. AIR FLASK
30. RELIEF VALVE
31. PLUG ASSEMBLY
32. RETAINING-RING SEGMENT
33. RUBBER GLAND AND GLAND NUT

Figure 2-11. Air Flask, Water Compartment, and Fuel Tank Assembly.

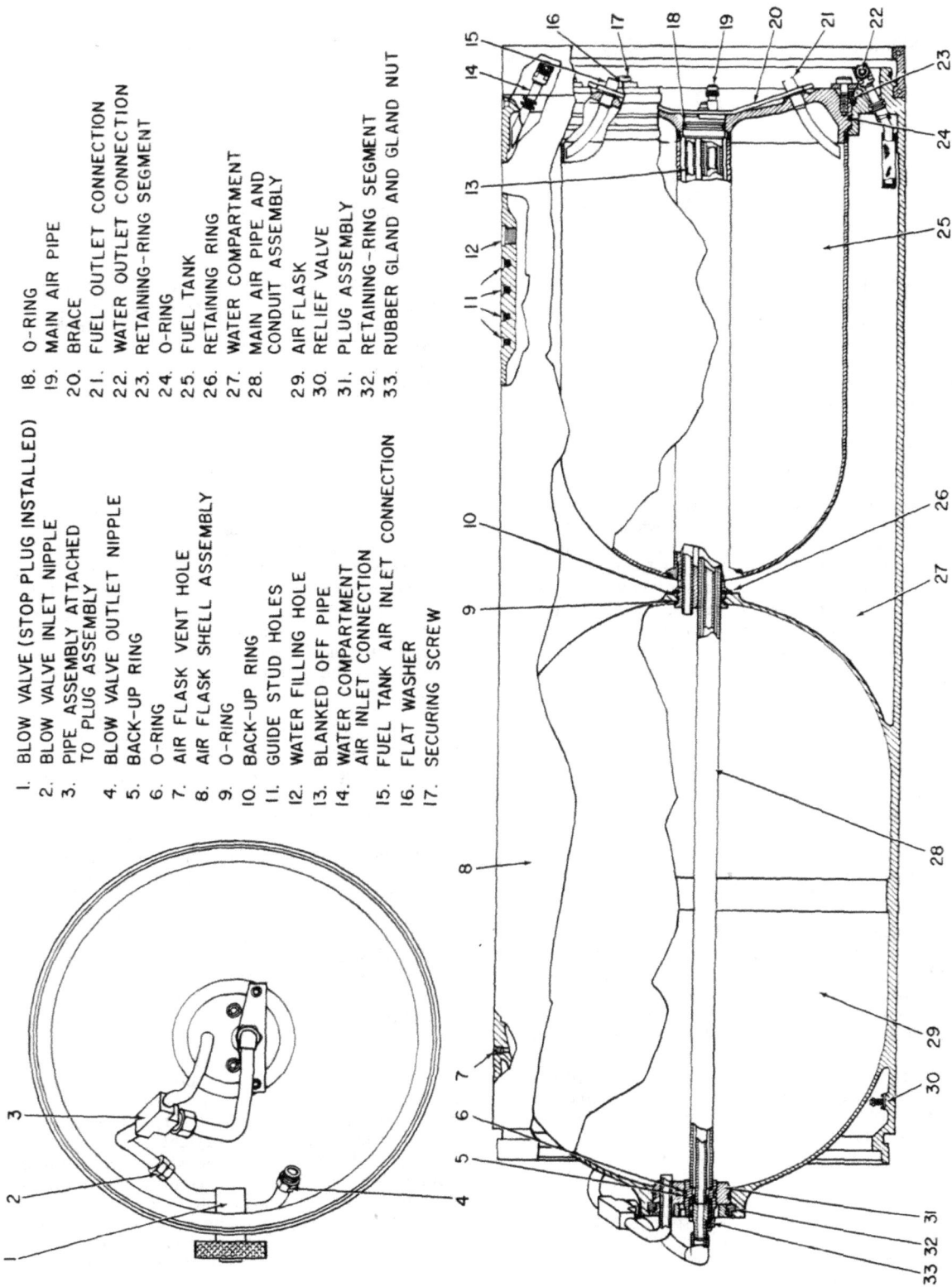

Figure 2-12. Main Air Pipe and Conduit Assembly.

plug seals the aft end of the central tube of the fuel tank. The main-air pipe passes through the tube of the conduit. The forward end of the main-air pipe is sealed in the conduit tube by a rubber gland and gland nut and is connected by a pipe assembly to the tee-connector of another pipe assembly that interconnects the air flask and blow valve. The aft end of the main-air pipe is connected to its mating main-air pipe of the plugs and pipes assembly in the Navol tank and valve compartment.

2-34. Square machined surfaces at each end of the main-air pipe accommodate a wrench when the pipe is being connected and disconnected and accommodate a bracket and brace which keep it from turning after its installation. The square surfaces fit into square slots in the bracket and brace. The bracket is secured to the forward face of the air flask forward bulkhead. The brace is secured to the after face of the fuel tank after bulkhead. When the main-air pipe is disconnected from the pipe assembly at its forward end, the gland nut is loosened, and the bracket and brace are removed; it can be moved 8 to 10 inches back from the conduit. This allows the aft end of the main-air pipe to be connected and disconnected at the plug and pipes assembly of the Navol tank and valve compartment.

2-35. Bulkhead Assembly. The bulkhead assembly has a plug that seals off the center of the forward dome-shaped bulkhead of the air flask (figure 2-11). The plug is equipped with piping to carry air-flask pressure to the forward end of the main-air pipe and conduit assembly and the blow valve. Two O-rings and two backup rings are installed on the plug to seal off the forward dome-shaped bulkhead of the air flask. Three retaining-ring sections, a clamp-plate, and a snapring hold the bulkhead assembly in place. The larger O-ring seals the bulkhead assembly in the dome-shaped bulkhead. The smaller O-ring seals the conduit of the main-air pipe and conduit assembly in the bulkhead assembly. The backup rings support the two O-rings for added insurance in retaining high-pressure air in the air flask.

2-36. Blow Valve Assembly. The blow valve assembly (figure 2-13) is a manually operated, fully opened or closed valve. It is installed on the starboard side of the forward flask shell skirt (figure

2-11), secured to the shell by a holding screw and a small setscrew. Its inlet connection receives air-flask pressure through the piping of the bulkhead assembly. When the valve is closed, a warhead or exercise head can be separated from or assembled to the energy section without bleeding the air flask. When a warhead is used, a flush-type plug is installed in the top of the blow valve and a blanking cap is installed on the outlet connection of the blow valve to safely seal off the air-flask pressure not required for warhead operation. When the exercise head is used, a stop plug is installed in the top of the blow valve; it is removed only for tests and just before loading the torpedo in the tube. A pipe assembly connects the outlet connection of the blow valve to the air-release mechanism in the exercise head. Upon removal of the stop plug, air-flask pressure is banked at the air-release mechanism. O-rings seal the valve stem in the holding screw, seal the holding screw in the valve body, and seal the holding screw.

2-37. Water Compartment Air-Inlet Connection. The air-inlet connection is installed in the top of the after flanged end of the water compartment portion of the flask shell assembly (figure 2-11). It is fitted with an O-ring to provide a watertight and airtight joint in the after flanged end of the water compartment. A spring pin provides alinement. A retaining plate secured to two studs with nuts holds the air-inlet connection in place. The air-inlet connection is equipped with a quick-connect fitting, color-coded black. The fitting accepts a flexible pipe assembly that carries working air pressure or venting pressure, depending on operation of the water-air check and vent valve installed in the Navol tank and valve compartment.

2-38. Water-Outlet Connection. The water-outlet connection is installed in the bottom of the after flanged end of the water compartment portion of the flask shell assembly (figure 2-11). It is fitted with an O-ring to provide a watertight and airtight joint in the after flanged end of the water compartment. A spring pin provides alinement. A retaining plate secured to two studs with nuts holds the water-outlet connection in place. The water-outlet connection is equipped with a quick-connect fitting, color-coded green. The fitting accepts a flexible pipe assembly that carries water from the water

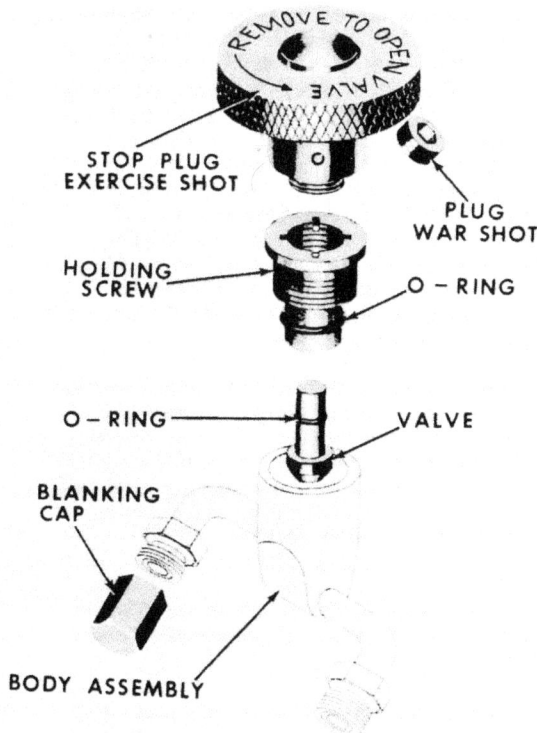

Figure 2-13. Blow Valve Assembly, Exploded View.

STOP PLUG EXERCISE SHOT
PLUG WAR SHOT
HOLDING SCREW
O-RING
O-RING
VALVE
BLANKING CAP
BODY ASSEMBLY

compartment to the water delivery valve in the Navol tank and valve compartment. A screen strainer is attached to the forward end of the water-outlet connection to prevent foreign matter in the water compartment from entering the water delivery system.

2-39. Relief Valve Assembly. The relief valve assembly is installed in the forward bottom skirt of the flask shell assembly (figure 2-11). It consists of a flanged body, a valve stem, a washer, a coil spring, a spring retainer, and a locknut. The flanged body and valve stem have flat valve seats sealed by the washer when the valve stem is under tension of the coil spring. The spring tension is maintained by the spring retainer and is secured by the locknut. The flanged body is riveted to the flask shell. The relief valve is designed to relieve any pressure of approximately 2.5 lbs or greater that may build up between the warhead bulkhead or into the exercise head because of leakage at a high-pressure air connection.

2-40. Guide Studs. There are two guide studs mounted on the top centerline and near the aft end of the air flask, water compartment, and fuel tank assembly (figures 1-3 and 2-11). The forwardmost guide stud is the longer one, secured by four screws, and is called the flat guide stud. The other one is called the auxiliary guide stud and is secured by one

screw. The longer, or flat, guide stud rides in a groove of the torpedo firing tube to aline the torpedo; it butts against the tube stop bolt when the torpedo is completely loaded in the tube. The auxiliary guide stud maintains torpedo alinement when the flat guide stud passes a void portion of the firing tube groove as the torpedo is being ejected from the tube.

2-41. Fuel Tank Assembly. The fuel tank assembly (figure 2-14) is installed in the after flanged bulkhead of the water compartment portion of the air flask, water compartment, and fuel tank assembly (figure 2-11). The outer shell of the fuel tank has a spherical forward end, a central tube that provides a passage for the main-air pipe and conduit assembly, and a dome-shaped after bulkhead that is capable of supporting the entire weight of the installed and loaded fuel tank. When the fuel tank is installed, its exterior surfaces complete the formation of the water compartment.

2-42. The cylindrical portion, central tube, and bulkhead are welded together to form an integral unit. An air-inlet connection, color-coded white, is mounted on the vertical centerline near the top of the bulkhead; it has a quick-connect fitting that accepts a flexible pipe assembly. A fuel-outlet connection is mounted on the vertical centerline near the bottom of the bulkhead. This connection has a quick-connect fitting, color-coded yellow, that accepts a flexible pipe assembly. The air-inlet connection and flexible pipe assembly carry working air pressure or venting pressure, depending on operation of the fuel-air check and vent valve installed in the Navol tank and valve compartment. The fuel-outlet connection and the flexible pipe assembly carry fuel from the fuel tank to the fuel delivery valve in the Navol tank and valve compartment. The fuel tank is filled through a fuel-filling hole and associated piping of the Navol tank and valve compartment (figure 2-17), and through a flexible pipe and fuel-outlet connection of the fuel tank assembly.

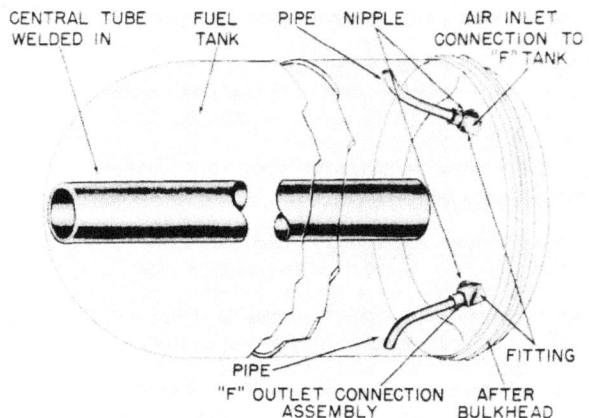

Figure 2-14. Fuel Tank Assembly.

2-43. Lock-Ring Segments and Brace. When the fuel tank is installed, an O-ring is placed in a groove around the perimeter at the afterend of the fuel tank (figure 2-11) to seal off the after flanged bulkhead of the water compartment. The fuel tank is secured in place by three lockring segments secured by 12 retaining washers and capscrews. A locating pin on the top centerline of the after flanged bulkhead of the water compartment fits into a slot in the after perimeter of the fuel tank to insure that the fuel tank is installed in its upright position. A brace, two lockwashers, and two capscrews secure the aft end of the main-air pipe and conduit assembly in the after bulkhead of the fuel tank.

2-44. NAVOL TANK AND VALVE COMPARTMENT. The Navol tank and valve compartment (figure 2-15) stores Navol, distributes air and fluids from start to end of a torpedo run, and monitors the decomposition rate of the Navol. When the Navol tank and valve compartment is received from the FIR activity, it is loaded with Navol; contains the fuel and water sprays, Navol restriction, and exhaust valve assemblies which are FIR parts used in the afterbody; and requires little preparation for torpedo use. As its name implies, it consists mainly of the Navol tank itself and a valve compartment. The valve compartment is divided into wet and dry sections, separated by a bulkhead.

2-45. Navol Tank. The Navol tank is aluminum alloy, is compatible with Navol, and has suitable strength to withstand torpedo working air pressure. It is formed from a weldment of a cylindrical outer shell having a dome-shaped bulkhead at each end, a tube through its center, and internal baffle plates. A steel cylindrical shell is shrunk over the aluminum weldment and extends slightly forward to provide a short skirt and joint ring for attachment of the air flask, water compartment, and fuel tank assembly. The steel outer shell also has an aft skirt that has considerably more length to form the exterior of the wet and dry sections of the valve compartment and has another joint ring for attachment of the afterbody. Steel stops are welded to the steel outer shell to lock the Navol tank shell in place. The internal baffle plates provide additional reinforcement and prevent rapid shifting of Navol that would otherwise occur when there is a change in torpedo attitude during a run. (Such shifting would affect torpedo trim characteristics and cause erratic performance or shutdown of the torpedo.) An air inlet and vent fitting is mounted in the top center of the after bulkhead of the Navol tank. A Navol outlet and fill fitting is mounted in the bottom center of the after bulkhead. The central tube of the Navol tank houses the plug and pipes assembly.

2-46. Plug and Pipes Assembly. The plug and pipes assembly (figure 2-16) consists of five pipe assemblies held in place by a plug at the after pipe ends and by two supports along the pipe lengths. The after pipe ends are brazed into holes of the plug and protrude from the afterface of the plug. The surfaces of the protruding pipe ends and afterface of the plug have a fine finish for acceptance of

O-rings. Upon assembly of the Navol tank and valve compartment, the pipe ends and plug afterface are fitted with O-rings and a manifold to seal off the afterend of the Navol tank central tube. The manifold has pipe fittings to accept pipe assemblies installed on the frame, valves, and pipe assembly in the dry section. The two supports have an inner and outer portion made of rubber. Each portion contains half-round holes with diameters equal to outside diameters of the pipes. A central hole in the inner portion of each support has a metallic ring molded in the rubber. These holes, another hole in the plug, and mating ports of the manifold allow venting of the Navol tank central tube in the event of a pipe failure. The inner and outer portions of the supports are held tightly in place by metallic straps wrapped around the outer portions.

2-47. Each pipe is given a descriptive name, serves a specific purpose, and all but one is color-coded and has a threaded elbow attached at its forward end to accept a flexible pipe assembly. The other end of each flexible pipe assembly is equipped with a quick-connect fitting. Each pipe of the plug and pipes assembly is briefly described as follows:

2-48. Main-Air Pipe. This pipe interconnects the extendible main-air pipe of the air flask and main-air port of the manifold. It is not color-coded and has a duoseal fitting.

2-49. Fuel Filling and Outlet Pipe. This pipe, together with its attached flexible pipe, interconnects the filling and outlet port of the manifold. It is color-coded yellow.

2-50. Fuel-Air Inlet Pipe. This pipe, together with its attached flexible pipe, interconnects the working pressure air-inlet connection of the fuel tank and working-pressure air port of the manifold. It is color-coded white.

2-51. Water-Outlet Pipe. This pipe, together with its attached flexible pipe, interconnects the water-outlet connection of the water compartment and water delivery port of the manifold. It is color-coded green.

2-52. Water-Air Inlet Pipe. This pipe, together with its attached flexible pipe, interconnects the working pressure air-inlet connection of the water compartment and working-pressure air port of the manifold. It is color-coded black.

2-53. Valve Compartment. The valve compartment houses the various valves, pressure switches, piping, and power pack to control the amount and sequential delivery of air, Navol, water, and fuel from start to finish of the torpedo run. It also contains a stop and charging valve for charging and bleeding the air flask, facilities for filling and draining the Navol and fuel tanks, and a monitoring unit to determine the decomposing rate of Navol. An anti-circling-run device causes a torpedo shutdown before the torpedo could circle and strike the firing craft in the event of a torpedo malfunction. A

1. FLEXIBLE PIPE TO FUEL TANK AIR-INLET CONNECTION
2. FLEXIBLE PIPE TO WATER COMPARTMENT AIR-INLET CONNECTION
3. QUICK-CONNECT COUPLINGS (4)
4. FLEXIBLE PIPE TO FUEL TANK WATER-OUTLET CONNECTION
5. FLEXIBLE PIPE TO WATER COMPARTMENT WATER-OUTLET CONNECTION
6. MAIN AIR PIPE DUO-SEAL CONNECTION
7. PLUG AND PIPE ASSEMBLY
8. NAVOL TANK
9. NAVOL VENT PIPE
10. NAVOL DELIVERY PIPE
11. GYRO INITIAL SPIN (HIGH-PRESSURE AIR) DELIVERY
12. MONITORING UNIT
13. POT PRESSURE PIPE
14. IGNITER PIPES
15. WORKING AIR PRESSURE DELIVERY (FLEXIBLE) PIPE
16. ACCESS OPENING
17. MONITORING UNIT PROBE CABLE
18. FUEL DELIVERY PIPE
19. NAVOL SURVEILLANCE AND POWER CABLE
20. TURBINE BULKHEAD GLAND FITTING
21. WET SECTION
22. DRY SECTION BULKHEAD
23. O-RING
24. WATER DELIVERY PIPE
25. DRY SECTION
26. POWER PACK
27. BAFFLES

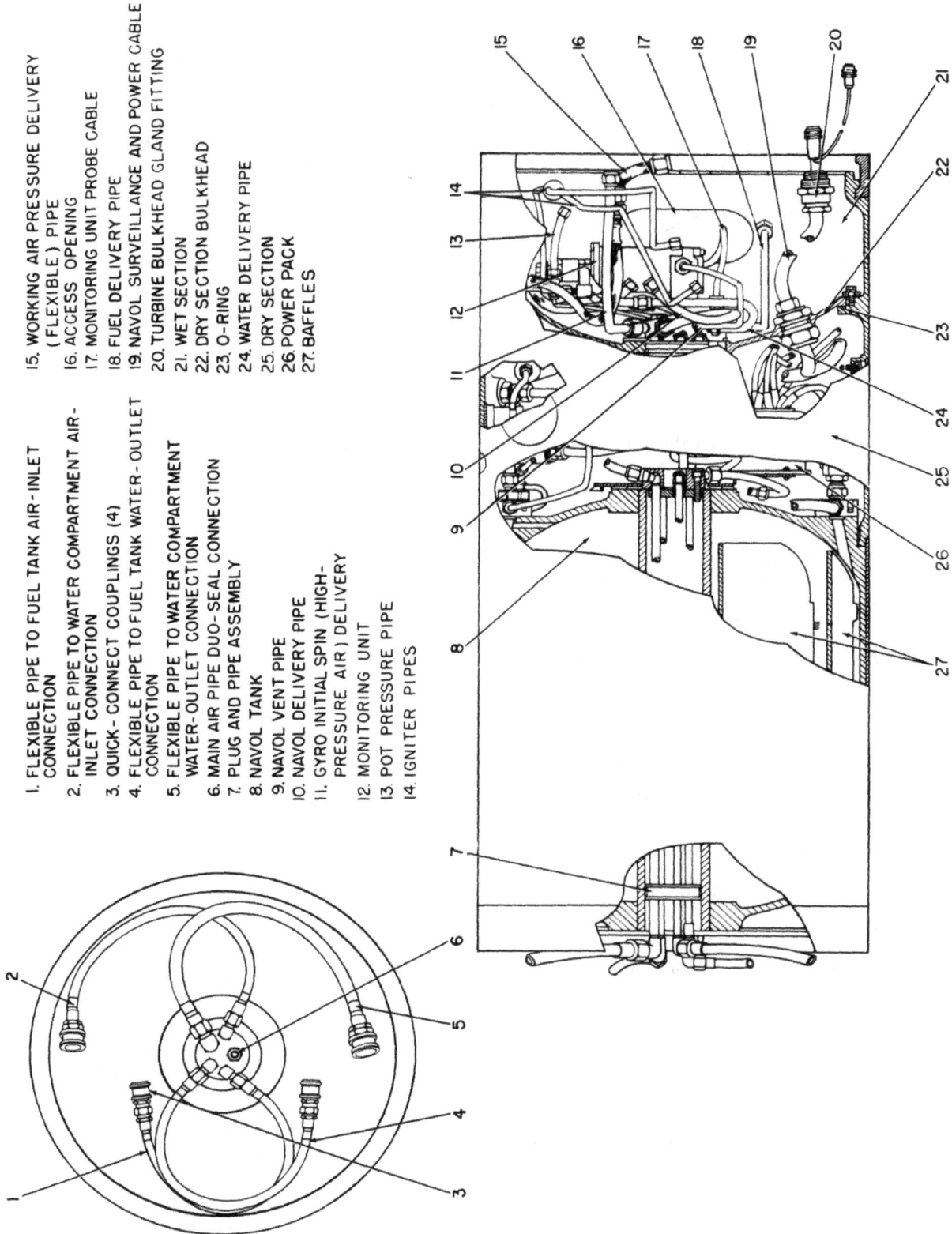

Figure 2-15. Navol Tank and Valve Compartment Assembly.

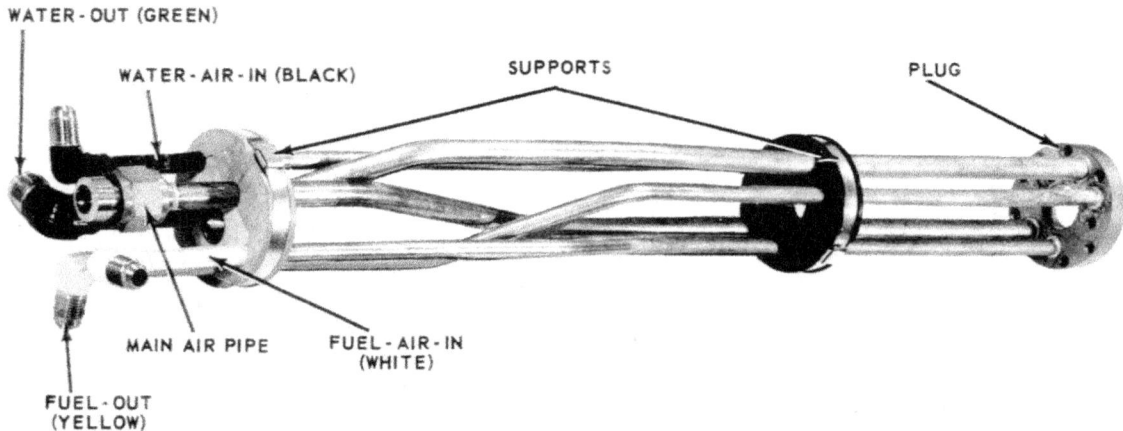

Figure 2-16. Plug and Pipes Assembly.

sea-pressure switch prevents premature torpedo firing. The valve compartment is composed of the dry section (figure 2-17), which is the forwardmost section, and the wet section (figure 2-18). A bulkhead seals off the aft end of the dry section and accommodates a plug, pipes, and manifold assembly, passage of a Navol surveillance and power cable, and passage of a monitoring unit cable. The wet section contains pipe assemblies that interconnect the related components of the energy section and afterbody, houses the monitoring unit, and provides space for the combustion system of the afterbody. The wet section is so called because it has several openings to allow entry of sea water for cooling the combustion system during a torpedo run. These openings also provide an escape for the gases expelled from the monitoring unit. The dry section contains the remaining major components of the Navol tank and valve compartment, as follows:

1. Frame Assembly.

2. Manifold on Frame Assembly.

3. Stop and Charging Valve.

4. Navol Delivery and Shutdown Valve.

5. Air, Fuel, and Water Delivery Valves.

6. Navol, Water, and Fuel Air Check and Vent Valves.

7. Air Strainer Assembly.

8. Pressure Reducing Valve, Manifold, and Rupture Disc Assembly.

9. Navol Pressure Switch.

10. Pot Pressure-Sensing Switch.

11. Anti-Circling-Run Device.

12. Power Pack and Test Switch Assembly.

13. Filling and Vent Flanges.

14. Sea Water Pressure Switch.

2-54. Dry Section Bulkhead. The dome-shaped bulkhead between the wet and dry sections (figure 2-18) seats on a bearing surface of a ring-shaped flange, which is an integral part of the aftershell skirt of the Navol tank and valve compartment. It is secured to the flange with 20 clamps and screws with its concave side facing aft. An O-ring installed between the bulkhead and flange provides a watertight joint. A hole in the center of the bulkhead accommodates the plug, pipes, and manifold assembly. Two holes in the lower right quadrant of the bulkhead accept fittings: one for sealing the passage of the Navol surveillance and power cable, the other for sealing the passage of the monitoring unit cable. A bracket, mounted adjacent to the center hole, is secured with two screws and supports the monitoring unit. Another bracket, held by two of the bulkhead-securing screws at the upper right quadrant of the bulkhead, supports a fitting and the pot pressure-sensing pipe.

2-55. Plug, Pipes, and Manifold Assembly. The plug, pipes, and manifold assembly is made up of two units. One unit consists of the plug assembly (figure 2-19), which is installed on the dry side of the center hole in the wet-dry section bulkhead. The other unit consists of the wet section manifold, which is installed on the wet side of the bulkhead (figure 2-18). When both units are installed, an O-ring is installed on the plug to create a watertight seal between the plug and bulkhead hole; mating parts of the plug and manifold are sealed by O-rings; and both units are secured together by six screws. Shoulders on the plug and manifold

STOP AND CHARGING VALVE

WATER AIR CHECK AND VENT VALVE

PLUG ASSEMBLY

WATER DELIVERY VALVE

NAVOL PRESSURE SWITCH

FUEL AND WATER OVERBOARD VENT OPENING AND INLET TO SEA-WATER PRESSURE SWITCH

FUEL FILLING ACCESS OPENING

FUEL AIR CHECK AND VENT VALVE

FUEL VENT SCREW

POT PRESSURE SWITCH

FRAME AND VALVES ASSEMBLY

FUEL DELIVERY VALVE

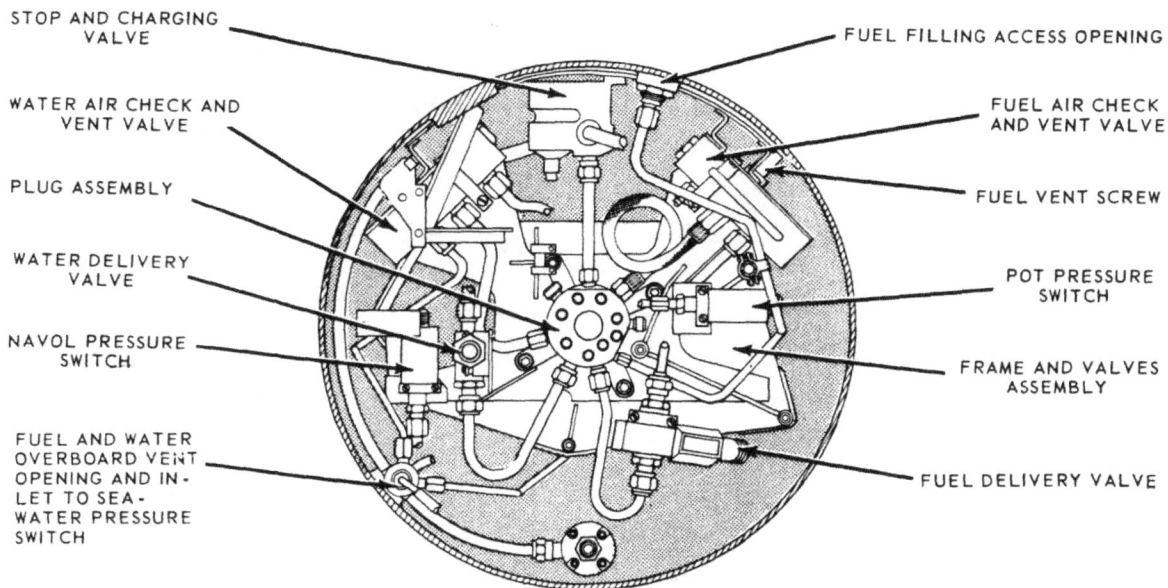

PARTIAL ASSEMBLY VIEW LOOKING FORWARD

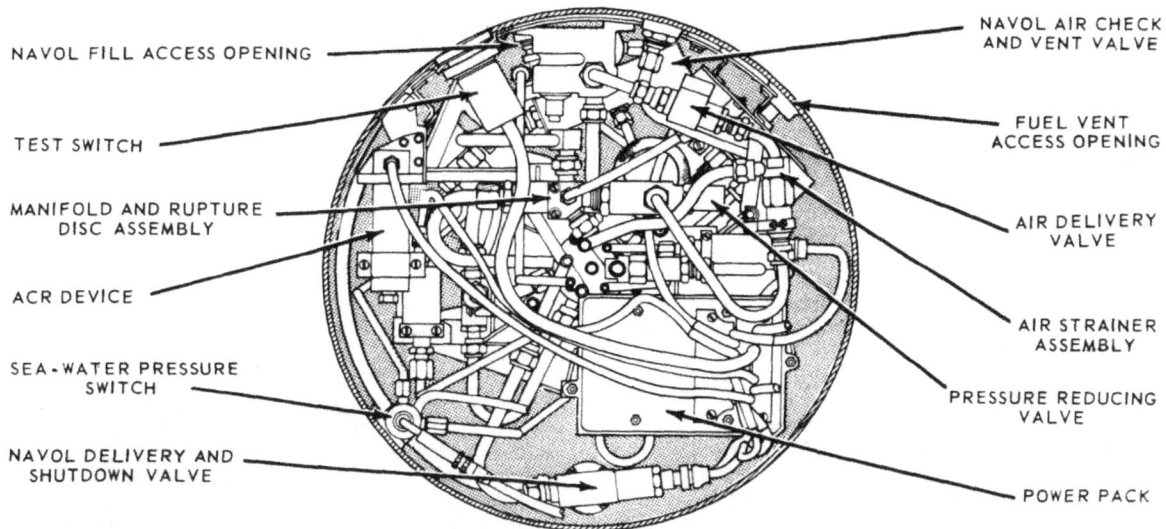

NAVOL FILL ACCESS OPENING

TEST SWITCH

MANIFOLD AND RUPTURE DISC ASSEMBLY

ACR DEVICE

SEA-WATER PRESSURE SWITCH

NAVOL DELIVERY AND SHUTDOWN VALVE

NAVOL AIR CHECK AND VENT VALVE

FUEL VENT ACCESS OPENING

AIR DELIVERY VALVE

AIR STRAINER ASSEMBLY

PRESSURE REDUCING VALVE

POWER PACK

COMPLETE ASSEMBLY VIEW LOOKING FORWARD

Figure 2-17. Dry Section of Navol Tank and Valve Compartment.

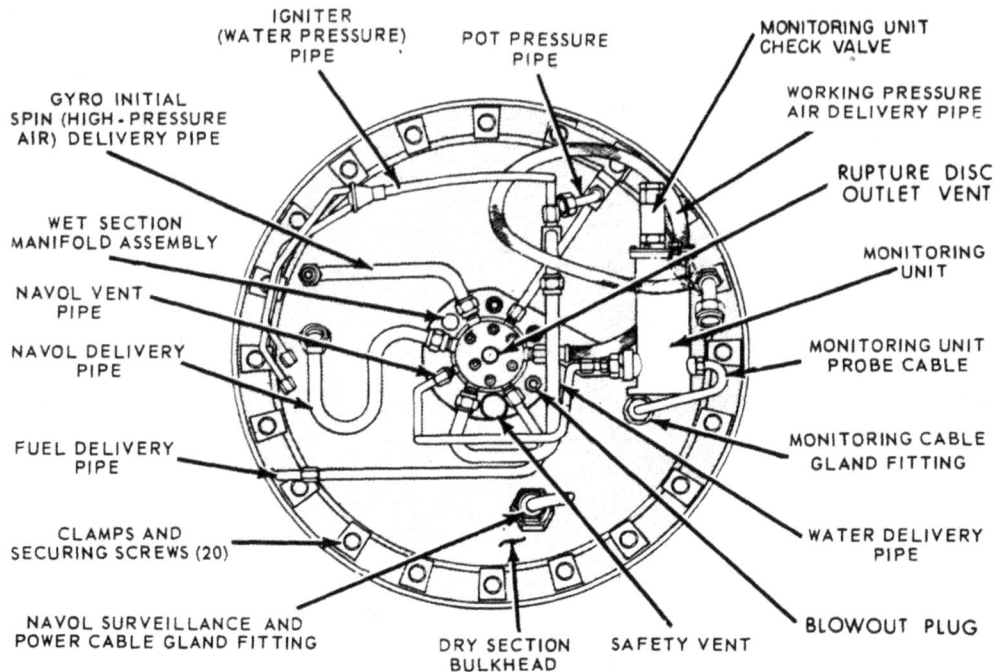

Figure 2-18. Wet Section of Navol Tank and Valve Compartment.

lock against the inner and outer edges of the bulkhead center hole, upon tightening the securing
screws. There are eight pipes attached to the forward face of the plug assembly. Two of the pipes
are welded to the plug, and the remaining six pipes
are silver-soldered. Each pipe protrudes slightly
from the aft face of the plug assembly to permit the
installation of O-rings and to enter portholes in the
wet section manifold. Passages through the manifold interconnect the pipes to nipples on the perimeter of the manifold. Each combination of pipe or
fitting, passage, and nipple serves a designated
purpose included in the delivery of high-pressure
air, low-pressure air, Navol, fuel, and water from
the energy section to the afterbody; delivery of
gases from the Navol tank to the monitoring unit;
delivery of combustion pot pressure to the pot-
pressure switch; and venting of high, abnormal reducer pressure from the dry section to the wet section and eventually overboard.

2-56. Manifold Pipe Assemblies. There are only
seven pipe assemblies attached to the nipples on the
wet section manifold in relation to the eight pipes
attached to the plug assembly. A hole in the center
of the manifold has no nipple or attached pipe assembly and serves as a vent opening for the high,
abnormal reducer pressure. The seven pipe assemblies are designated as follows (figure 2-18):

1. Pot Pressure Input.

2. Working Pressure Air Delivery.

3. Fuel Delivery.

4. Water Delivery.

5. Navol Vent.

6. Navol Delivery.

7. High-Pressure Air (Gyro Initial Spin) Delivery.

2-57. The pot-pressure input pipe is made up of
two pieces. One piece consists of a longer pipe
with one end connected to its manifold nipple and
the other end connected to a nipple of a fitting secured to a bracket held in place by two bulkhead
securing screws. The shorter pipe connects to
another nipple on the fitting and to a nipple on the
nozzle of the combustion system. This piping carries pressure that originates in the combustion
pot of the combustion system to related piping and
the pot-pressure switch in the dry section.

2-58. The flexible working-pressure air-delivery
pipe is connected to its manifold nipple and to the
low-pressure (working) air nipple on the turbine
bulkhead. It carries working-air pressure, orig-
inating at the pressure-reducing valve in the dry
section, to the turbine bulkhead nipple for afterbody distribution.

2-59. The fuel delivery pipe is connected to its
manifold nipple and to the nipple on the fuel spray

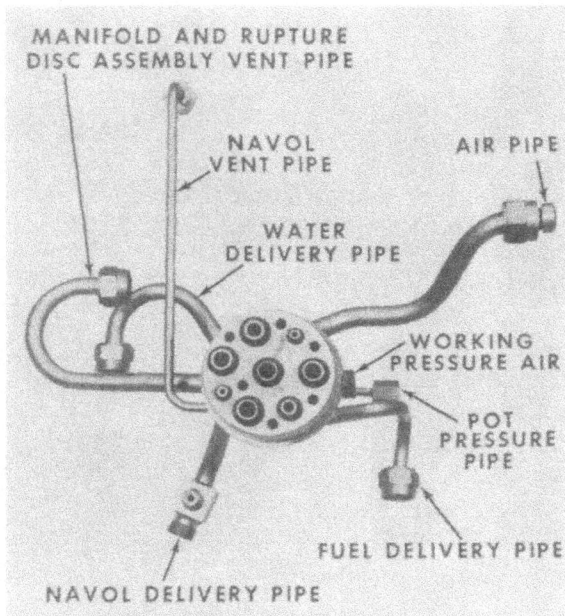

Figure 2-19. Plug Assembly.

assembly of the combustion system. It carries pressurized fuel from the fuel delivery valve in the dry section to the fuel spray assembly for combustion.

2-60. One part of the water delivery pipe connects to its manifold nipple and is silver-soldered into a tee-connector; another part is silver-soldered into the tee-connector and connects to the nipple on the water-spray assembly of the combustion system. It carries pressurized water from the water delivery valve in the dry section to the water spray assembly for cooling the combustion system. This water is eventually used to produce the driving steam.

2-61. A Y-shaped pipe assembly is attached to a nipple on the tee connector of the water-delivery piping and to the nipples on the igniters of the combustion system. Although this pipe assembly is not a functional part of the water-delivery piping, it carries the water-delivery pressure used to activate the igniters.

2-62. The Navol vent pipe is connected to its manifold nipple and to a nipple on the lower end of the monitoring unit. It carries gases of decomposing Navol to the monitoring unit.

2-63. The Navol delivery pipe is connected to its manifold nipple and to a nipple in the Navol restriction of the combustion system. It carries pressurized Navol from the Navol delivery valve in the dry section to the Navol restriction for supporting combustion. The Navol eventually produces additional driving steam.

2-64. The high-pressure (gyro initial-spin) air pipe is connected to its manifold nipple and to the high-pressure air nipple on the turbine bulkhead. It carries high-pressure air from the air delivery valve in the dry section to the turbine bulkhead nipple for distribution to the impulse valve of the spinning and unlocking mechanism, which gives the gyro its initial spin.

2-65. Except for the shorter pipe of the pot pressure-sensing piping, all pipe assemblies in the wet section of the Navol tank and valve compartment are fixed in position by the FIR activity. However, the tending activity must check the alinement of the pipe assemblies and test pipe connections for leakage.

2-66. Navol Monitoring Unit. The Navol monitoring unit (figure 2-20) is installed on a swivel bracket at the starboard side in the wet section. Its body consists of a flange assembly, a block, an inlet fitting, a probe receptacle boss, and a U-shaped tube fabricated to become a unit. A cap assembly, attached to the top of the body flange with securing screws, has two openings; one receives a plug, and the other receives a check valve. The plug is removed to fill the body compartment with a solution of distilled water and sodium nitrate. The one-way check valve allows gases to escape, while preventing entry of sea water. The receptacle boss receives the probe end of the monitoring cable. The inlet fitting receives the Navol vent pipe, which carries gases from the Navol tank. The body block has access holes for entry of the water and sodium nitrate solution and other access holes interconnected by the U-shaped tube. Both types of access holes terminate at a common chamber in the body block. The probe of the monitoring cable inserts into the common chamber.

2-67. Frame Assembly. The frame assembly is mounted in the dry section (figure 2-17). It is irregular in shape, consisting of a welded fabrication of a baseplate, extensions, a top plate, braces, and pads all made of steel. It mounts most of the dry section components.

2-68. Manifold on Frame Assembly. The manifold, located in the center of the frame assembly (figure 2-17), connects to the plug of the plug and pipes assembly. It consists of a ring-shaped body, eight nipples (two not used), and three radial brackets. Each nipple is connected by holes to bore openings in the aft face of the manifold body. The bore openings accept O-rings and pipe ends of the plug and pipes assembly. The eight nipples are welded to the manifold body. Six of them accommodate pipe assemblies for connections to the stop and charging valve; water and fuel air check and vent valves; fuel and water delivery valves; and fuel filling flange. Blanking caps are installed on the two unused nipples. The three brackets are welded to the manifold body and secure the manifold and frame to the Navol tank bulkhead with screws. Eight capscrews secure the manifold to the plug and pipes assembly.

Figure 2-20. Navol Monitoring Unit.

2-69. Stop and Charging Valve. The stop and charging valve (figure 2-21), located in the top center of the dry section (figure 2-17), is mounted on the top plate of the frame assembly. An opening in the dry section shell allows access to the valve. The opening has a cover to seal the dry section. The dual-type stop and charging valve is manually operated from outside the shell. A pipe assembly connects the inlet nipple of the stop and charging valve to the high-pressure air nipple of the manifold on the frame assembly; it carries pressure to and from the air flask. Another pipe assembly connects the outlet nipple of the stop and charging valve to the air delivery valve; it distributes air flask pressure when the torpedo is fired.

2-70. Navol Delivery and Shutdown Valve. The Navol delivery and shutdown valve (figure 2-22), located in the bottom center of the dry section (figure 2-17), is attached directly to the outlet fitting of the Navol tank. It is electrically initiated and squib activated. The outlet port of the valve is connected by a pipe assembly to the manifold in the center of the frame assembly. Other piping carries the Navol to the combustion system.

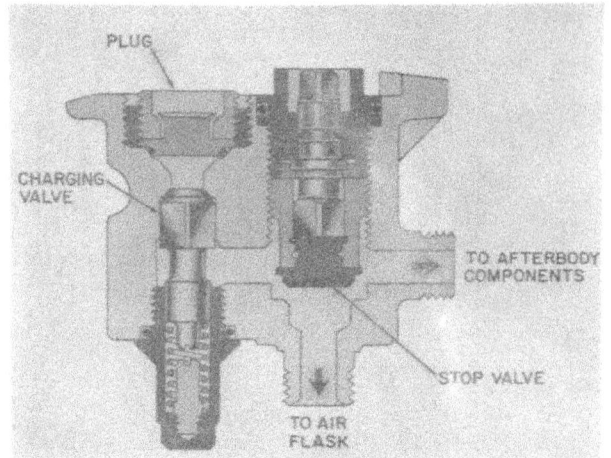

Figure 2-21. Stop and Charging Valve, Sectional View.

2-71. Air, Water, and Fuel Delivery Valves. The air, water, and fuel delivery valves are identical (figure 2-23). They are electrically initiated and squib activated. Each valve has two squibs to insure activation upon sequential operation of the power pack. Before activation, each valve is closed. The air delivery valve, mounted on the frame assembly, is located in the upper right quadrant of the dry section (figure 2-17). Its inlet port is connected by a pipe assembly to the outlet nipple of the stop and charging valve. Its outlet port is connected by piping to the inlet nipple of the air strainer and to the high-pressure (initial gyro-spin) air nipple of the plug and manifold assembly in the center of the wet-dry section bulkhead. The water delivery valve, mounted on the frame assembly, is located in the lower left quadrant of the dry section (figure 2-17). Its inlet port is connected by a pipe assembly to the water-outlet nipple of the manifold assembly in the center of the frame assembly. Its outlet port is connected by piping to the plug and manifold assembly in the center of the wet-dry section bulkhead. The fuel delivery valve, mounted on the frame assembly, is located in the lower right quadrant of the dry section (figure 2-17). Its inlet port is connected by a pipe assembly to the fuel-outlet nipple of the manifold assembly in the center of the frame assembly. Its outlet port is connected by piping to the plug and manifold assembly in the center of the wet-dry section bulkhead. A filter assembly in the outlet port prevents foreign material from restricting the normal flow of fuel into the combustion chamber.

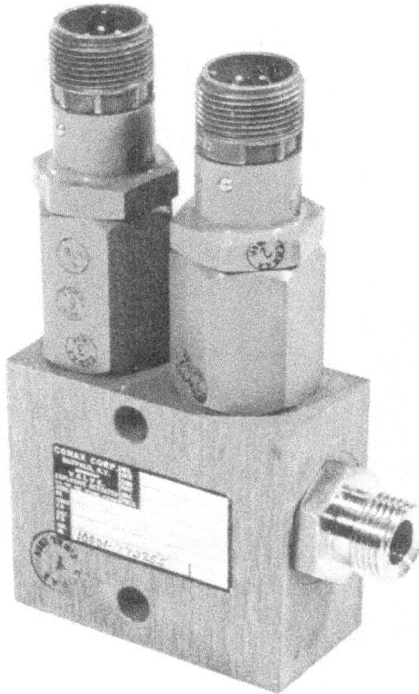

Figure 2-22. Navol Delivery and Shutdown Valve.

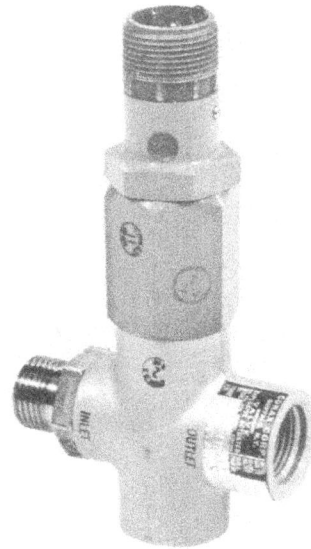

Figure 2-23. Typical Air, Water, and
Fuel Delivery Valve.

pressure air-outlet nipple is connected by piping to the air-inlet connection of the water compartment. Its venting-pressure outlet nipple is connected by a pipe assembly to the water and fuel overboard vent fitting.

2-75. The fuel air-check and vent valve (figure 2-26) is clamped to the frame assembly and is located in the upper right quadrant of the dry section (figure 2-17). Its working-pressure air-inlet nipple is connected by a pipe assembly to the manifold on the pressure-reducing valve. Its working-pressure air-outlet nipple is connected by piping to the air-inlet connection of the fuel tank. Its venting-pressure outlet nipple is connected by a pipe assembly to the water- and fuel-overboard vent fitting. It also has a vent screw, accessible upon removal of a cover plate from the dry section shell, that is used to manually vent the fuel tank.

2-76. Air Strainer Assembly. The air strainer assembly (figure 2-27) removes foreign matter from the delivery of air-flask (high-pressure) air. It is mounted in the upper right quadrant of the dry section (figure 2-17). A pipe fitting with a short length of pipe is silver-soldered to the strainer body and serves as an inlet connection. The short length of pipe is connected to the outlet nipple of the air delivery valve. The side nipple is connected to the plug and manifold assembly in the center of the wet-dry section bulkhead by a high-pressure (gyro initial-spin) air pipe assembly. Its outlet nipple is connected by a pipe assembly to the inlet nipple of the pressure-reducing valve.

2-72. Navol, Water, and Fuel Air-Check and Vent Valves. The Navol, water, and fuel air-check and vent valves consist mainly of a valve body, a spring-loaded air-check valve, a spring-loaded vent valve, and a vent valve piston.

2-73. The Navol air-check and vent valve (figure 2-24), mounted in a bracket, is located behind the frame assembly and in the upper center of the dry section (figure 2-17). Its working pressure air-inlet nipple is connected by a pipe assembly to the manifold of the pressure reducing valve. Its working pressure air-outlet nipple is connected by a pipe assembly to the rupture disk and screen assembly. Its venting pressure inlet nipple is connected by a pipe assembly to the filter, which is part of the Navol inlet fitting of the Navol tank. Its venting-pressure outlet nipple is connected by a pipe assembly to the monitoring unit.

2-74. The water air-check and vent valve (figure 2-25) is clamped to the frame assembly and is located in the upper left quadrant of the dry section (figure 2-17). Its working-pressure air-inlet nipple is connected by a pipe assembly to the manifold in the pressure-reducing valve. Its working

Figure 2-24. Navol Air-Check and Vent Valve.

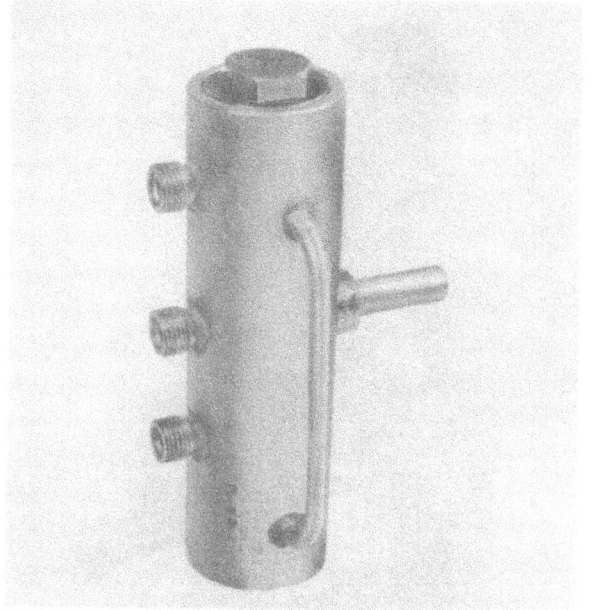

Figure 2-26. Fuel Air-Check and Vent Valve.

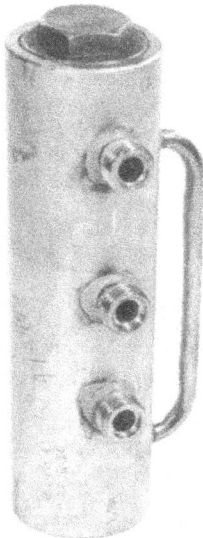

Figure 2-25. Water Air-Check and Vent Valve.

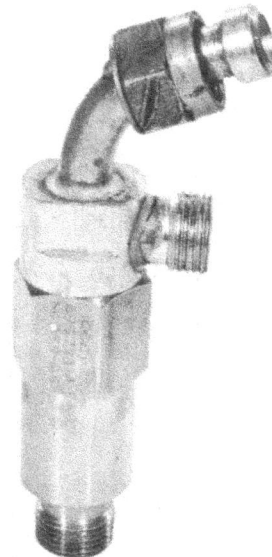

Figure 2-27. Air Strainer Assembly.

2-77. Pressure-Reducing Valve and Manifold and Rupture Disk Assembly. The pressure-reducing valve and manifold and rupture disk assembly (figure 2-28) are mounted in the upper center of the dry section (figure 2-17). The reducing valve reduces air-flask pressure to the working-air pressure (635 ±10 psi). The inlet port of the reducing valve is connected by a pipe assembly to the outlet nipple of the air strainer assembly. The manifold is threaded to the outlet port of the reducing valve and has four outlet nipples. The attached rupture disk assembly has one nipple. Three of the manifold outlet nipples are connected by pipe assemblies to the Navol, water, and fuel air-check and vent valves. The remaining manifold outlet nipple is connected by piping to the plug and manifold assembly in the center of the wet-dry section bulkhead for delivery of working-pressure air to the afterbody. The rupture disk, installed in the body of the rupture disk outlet port of the assembly, is ruptured in the event of a delivery of an abnormally high (1200 psi) working-air pressure from the reducing valve, thus preventing overpressurization of the Navol tank, water compartment, and fuel tank. The nipple of the rupture disk assembly is connected by piping to the plug and manifold assembly in the center of the dry section bulkhead to deliver the excessive pressure overboard.

2-78. Navol Pressure Switch. The Navol pressure switch (figure 2-29), mounted on the frame assembly, is located in the lower left quadrant of the dry section (figure 2-17). The open contacts of its switch are connected by a cable to the power pack. Its pressure-inlet nipple is connected to the piping that connects the Navol delivery and shutdown valve to the combustion system. When Navol is delivered, the switch actuates at 100 ±5 psi and completes the applicable power pack circuit to the fuel delivery valve.

2-79. Pot Pressure-Sensing Switch. Except for being set at a different operating pressure, the pot pressure-sensing switch is similar to the Navol pressure switch (figure 2-29). It is mounted on the frame assembly and located on the right of the horizontal centerline, in the upper right quadrant of the dry section (figure 2-17). The closed contacts of its switch are connected by a cable to the power pack. Its pressure-inlet nipple is connected by piping of the plug assembly in the center of the wet-dry section to the combustion system.

2-80. Anti-Circling-Run Device. The anti-circling-run device (ACR) (figure 2-30), mounted on a bracket attached to the frame assembly, is located in the upper left quadrant of the dry section (figure 2-17). It is composed of a manually wound, spring-powered gyro, a delay-on-dropout relay, and an explosive impact squib. It is connected to the power pack by a cable.

2-81. Power Pack and Test Switch Assembly. The power pack and test switch assembly (figure 2-31) is composed of timing circuits made up of four time-delay relays, four diodes, four resistors, and associated wiring mounted on terminal boards secured to a base plate. All components are sealed in potting compound and are contained in a metal case. Several cables protrude from a junction box in the case and are provided with connectors to complete the energy control circuit.

2-82. Filling and Vent Openings. The dry section shell has small openings to permit filling and venting the energy section. Near the top-right center of the shell is a fuel filling opening consisting of a threaded flange fitted with a plug and Teflon seat, connected by piping to the manifold assembly in the center of the frame assembly, and used in the filling of the fuel tank. At the upper right quadrant

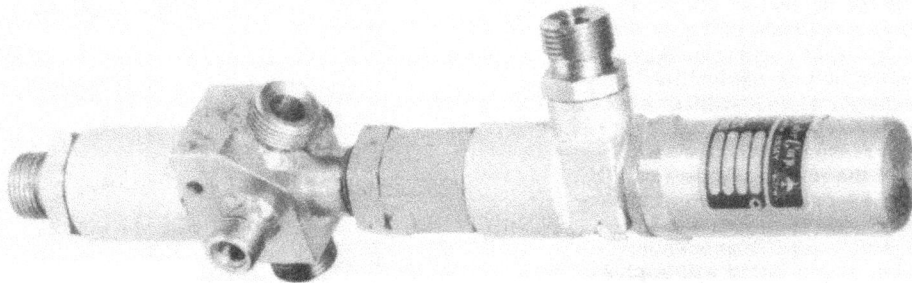

Figure 2-28. Pressure Reducing Valve and Manifold and Rupture Disk Assembly.

Figure 2-29. Typical Navol Pressure or Pot
Pressure-Sensing Switch.

Figure 2-30. Anti-Circling-Run Device.

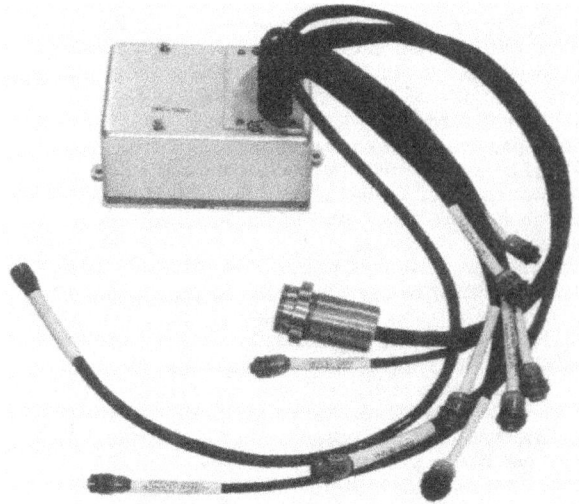

Figure 2-31. Power Pack and Test
Switch Assembly.

center of the Navol tank forward bulkhead, and used
to fill the Navol tank. Near the top-right center of
the shell is a Navol vent opening consisting of a
threaded flange fitted with a plug and Teflon seat,
connected by piping to the top center of the Navol
tank forward bulkhead, and used to vent the Navol
tank during the filling process. The Navol filling
and vent openings are also used during the emer-
gency stabilizing process.

2-83. AFTERBODY AND TAIL SECTION

2-84. The afterbody and tail section (figure 2-32)
is considered the main section of the torpedo. It
contains the major components (figure 2-33) required for torpedo propulsion; to preset and control the course, depth, and length of the straight
run; and to initiate a power shutdown in an emer-
gency. It consists of the following:

1. Afterbody Shell Assembly.
2. Combustion System.
3. Turbine and Gear Train Assembly.
4. Control Mechanism.
5. Idler Gear and Bracket Assembly.
6. Governor Assembly.
7. Deleted by CHANGE 2
8. Afterbody Electrical Cabling.
9. Afterbody Piping.
10. Afterbody Expendables.
11. Tail Assembly.

of the shell is a fuel vent opening consisting of a
flange equipped with a cover and O-ring which allow access to the vent screw of the fuel air-check
and vent valve for use in the manual venting of the
fuel tank. At the lower left quadrant of the shell is
a fuel and water-overboard vent opening consisting
of an open flange connected by pipe assemblies to
the fuel and water air-check and vent valves; it ex-
pels fuel and water into the sea at the end of a torpedo run. A sea-water pressure-activated switch
is mounted on the body of the fuel and water-overboard vent opening, which permits torpedo firing
only at a 10-foot or greater depth. Near the top-
left center of the shell is a Navol filling opening
that consists of a threaded flange fitted with a plug
and Teflon seat, connected by piping to the bottom

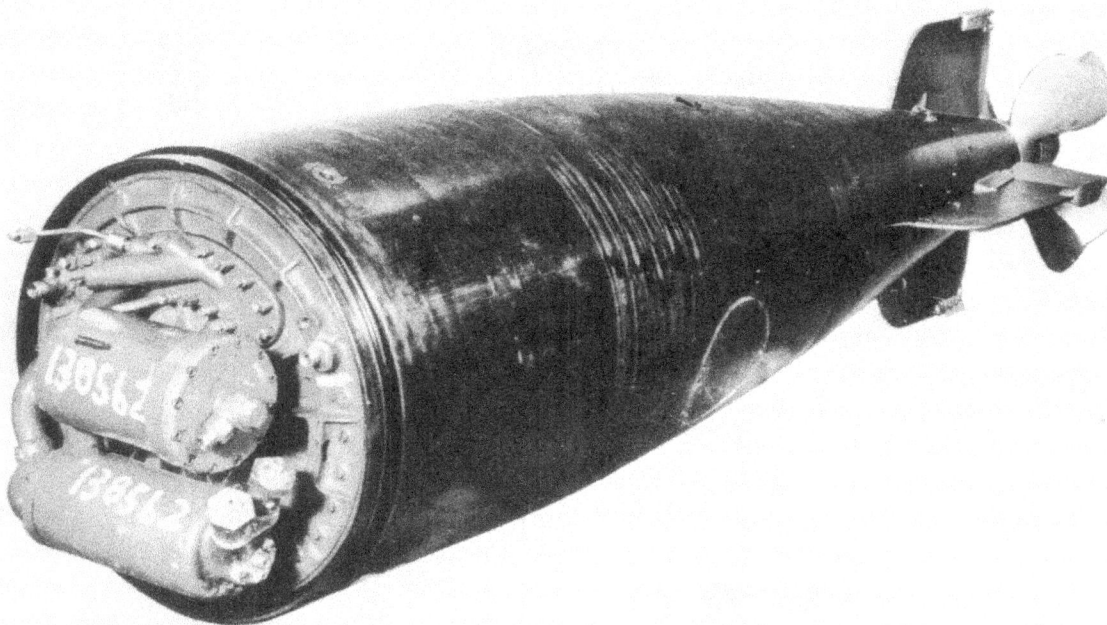

Figure 2-32. Afterbody and Tail Section.

2-85. AFTERBODY SHELL ASSEMBLY. The after-body shell assembly (figure 2-34) consists of the following major components:

1. Shell.

2. Joint Ring.

3. Strengthening Rings.

4. Flanges.

5. Vertical Bulkhead.

6. Exhaust System.

7. Handhole Covers.

8. Test Connection.

9. After Bulkhead Assembly.

10. Rudder Rod Assemblies.

2-86. Shell. The afterbody shell is made of sheet steel circularly formed so that its forward end conforms to the main diameter of the torpedo. Its contour is gradually curved and tapered, and the aft end conforms to the diameter at the forward end of the tail assembly. The shell exterior and interior surfaces are coated with air-drying phenolic resinoid material.

2-87. Joint Ring. The steel joint ring is a portion of the wedge-type joint used in assembly of the afterbody and tail section to the energy section; but no O-ring is used because watertightness is unnecessary between these two sections. However, it is welded completely around the circumference of the shell forward end to create a watertight seal for the afterbody interior and to provide structural strength for the shell forward end.

2-88. Strengthening Rings. Several steel rings are spotwelded to the shell inner surface, spaced to provide added structural strength to keep the shell rigid.

2-89. Flanges. There are several steel flanges of various sizes and shapes welded to the internal perimeters of many holes scattered on the shell contour. Some of the flanges, one for mounting the test connection and others for filling and draining the engine casing and afterbody, contain plugs and O-rings when not in use. Each of the remaining flanges is used for individual installation of the control mechanism, handhole covers, sea water inlet scoops, main connector receptacle of afterbody electrical cabling, and governor assembly.

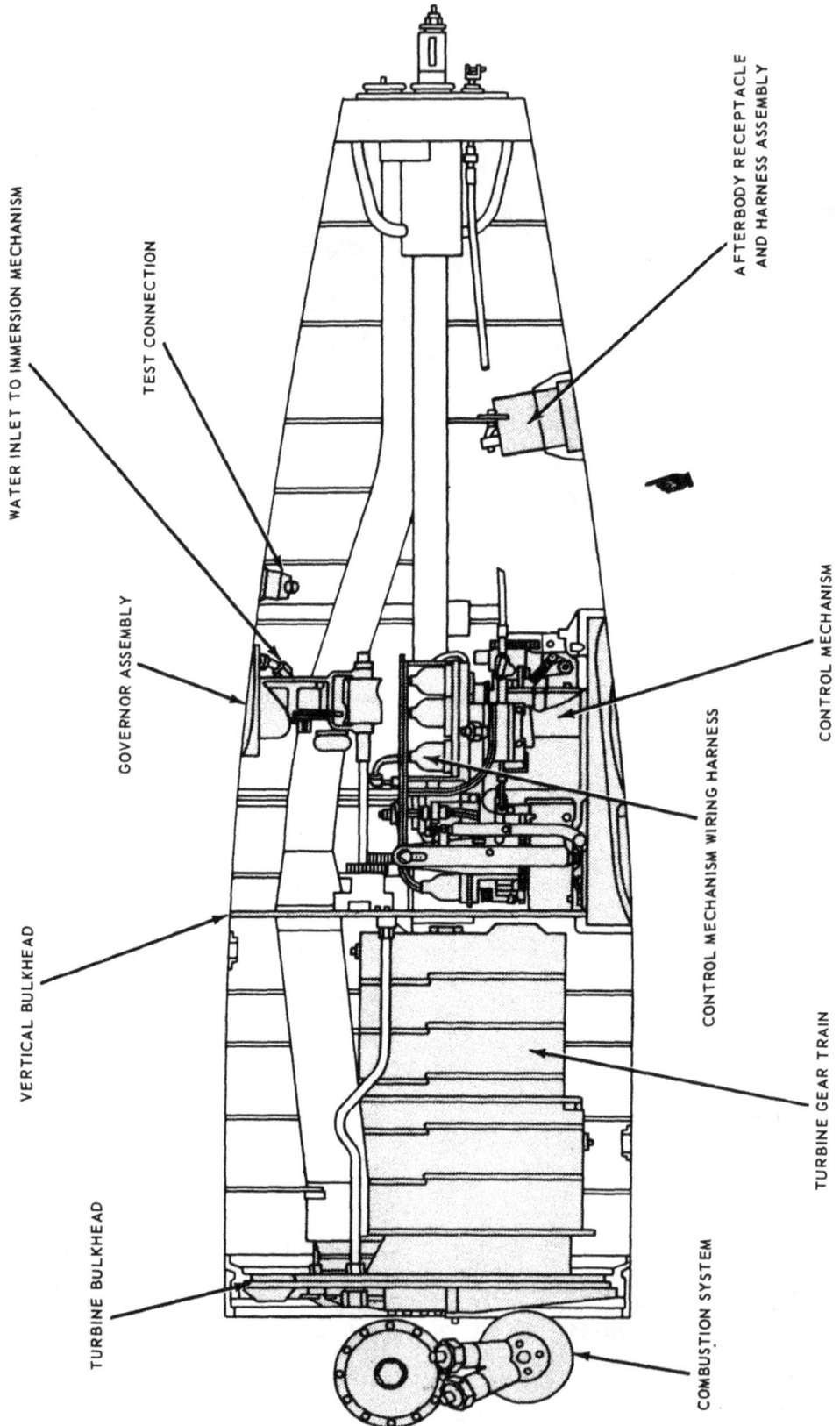

Figure 2-33. Afterbody.

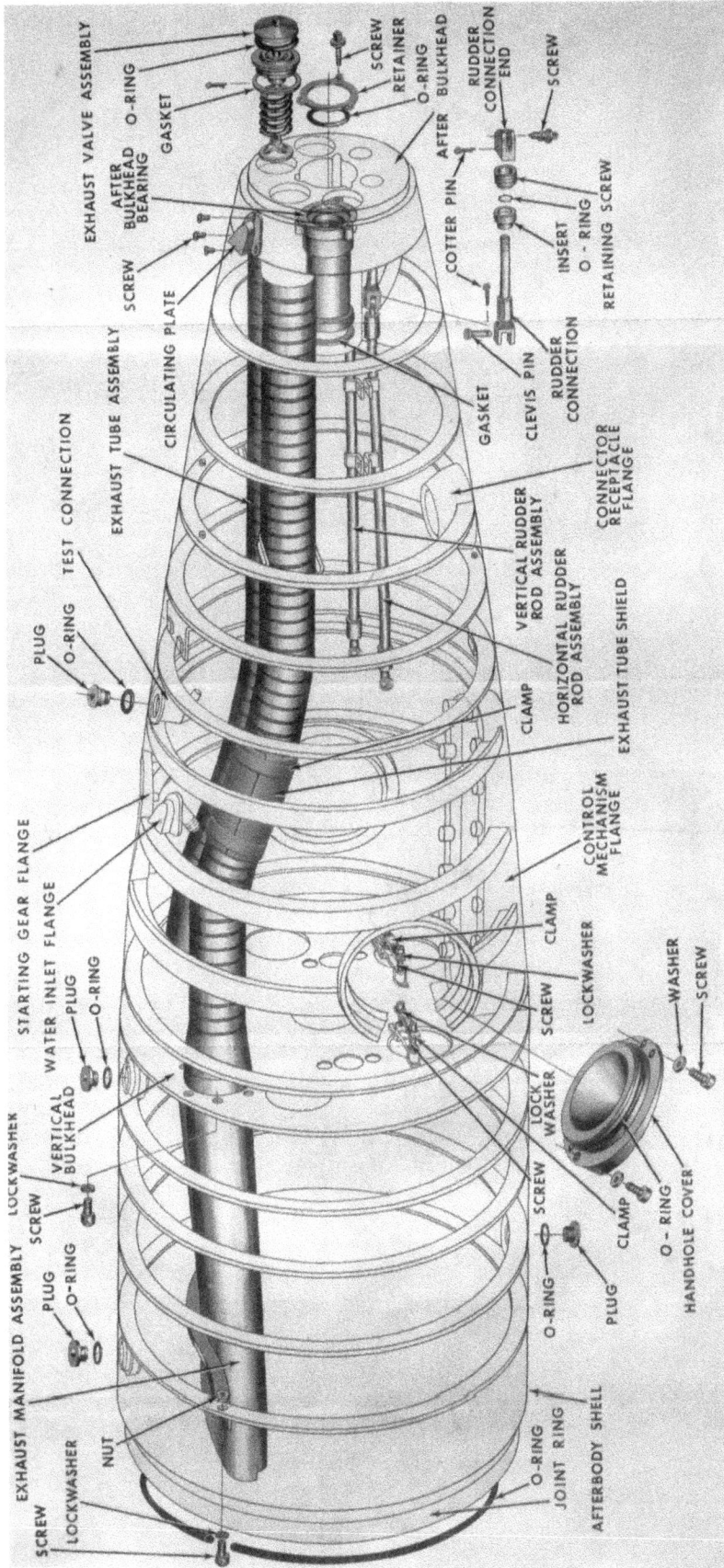

Figure 2-9-4. Afterbody Shell Assembly.

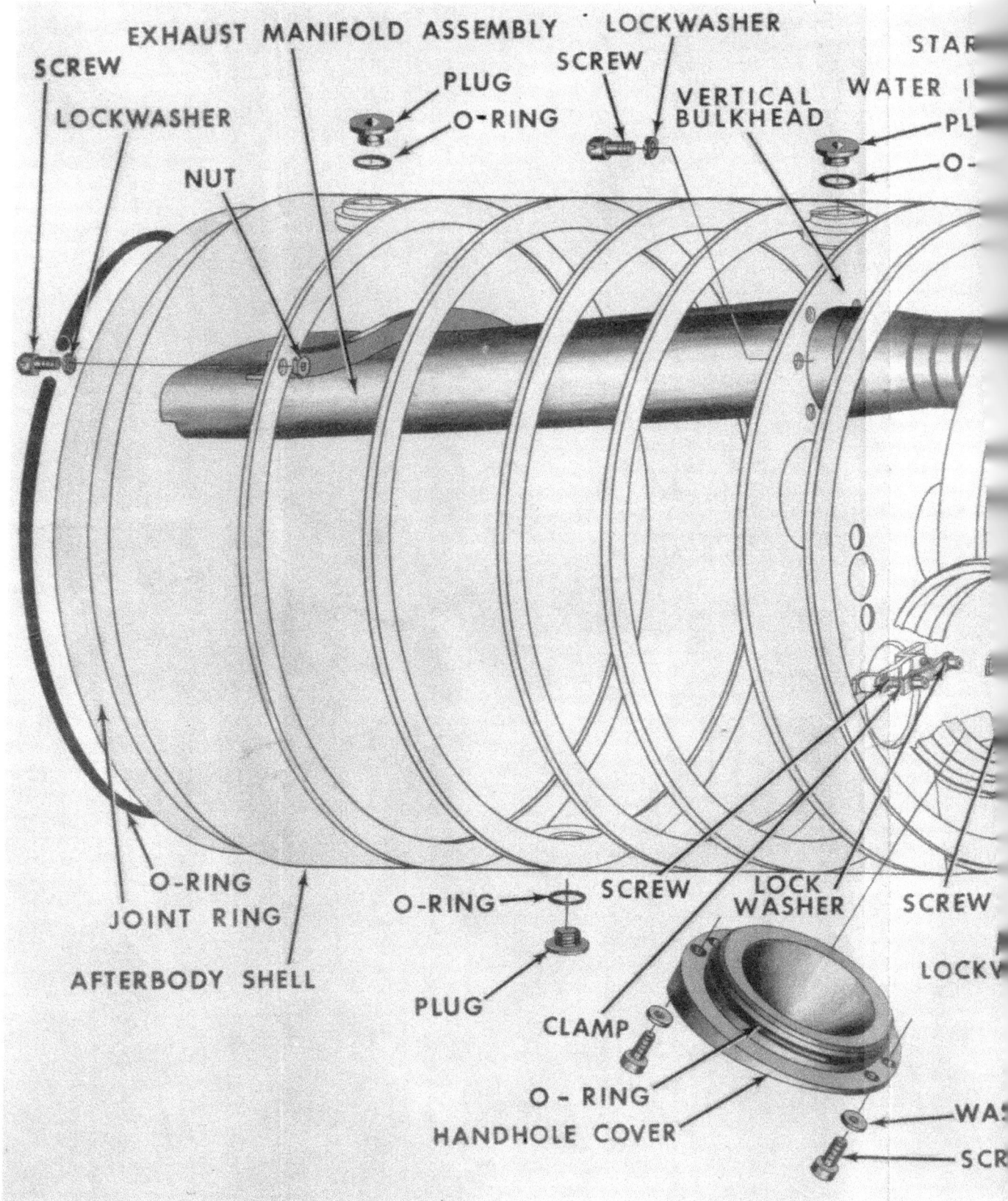

SCREW

LOCKWASHER

NUT

EXHAUST MANIFOLD ASSEMBLY

PLUG

O-RING

LOCKWASHER

SCREW

VERTICAL BULKHEAD

STAR

WATER I

PL

O-

O-RING

JOINT RING

AFTERBODY SHELL

O-RING

PLUG

SCREW

CLAMP

O-RING

HANDHOLE COVER

LOCK WASHER

SCREW

LOCKW

WAS

SCR

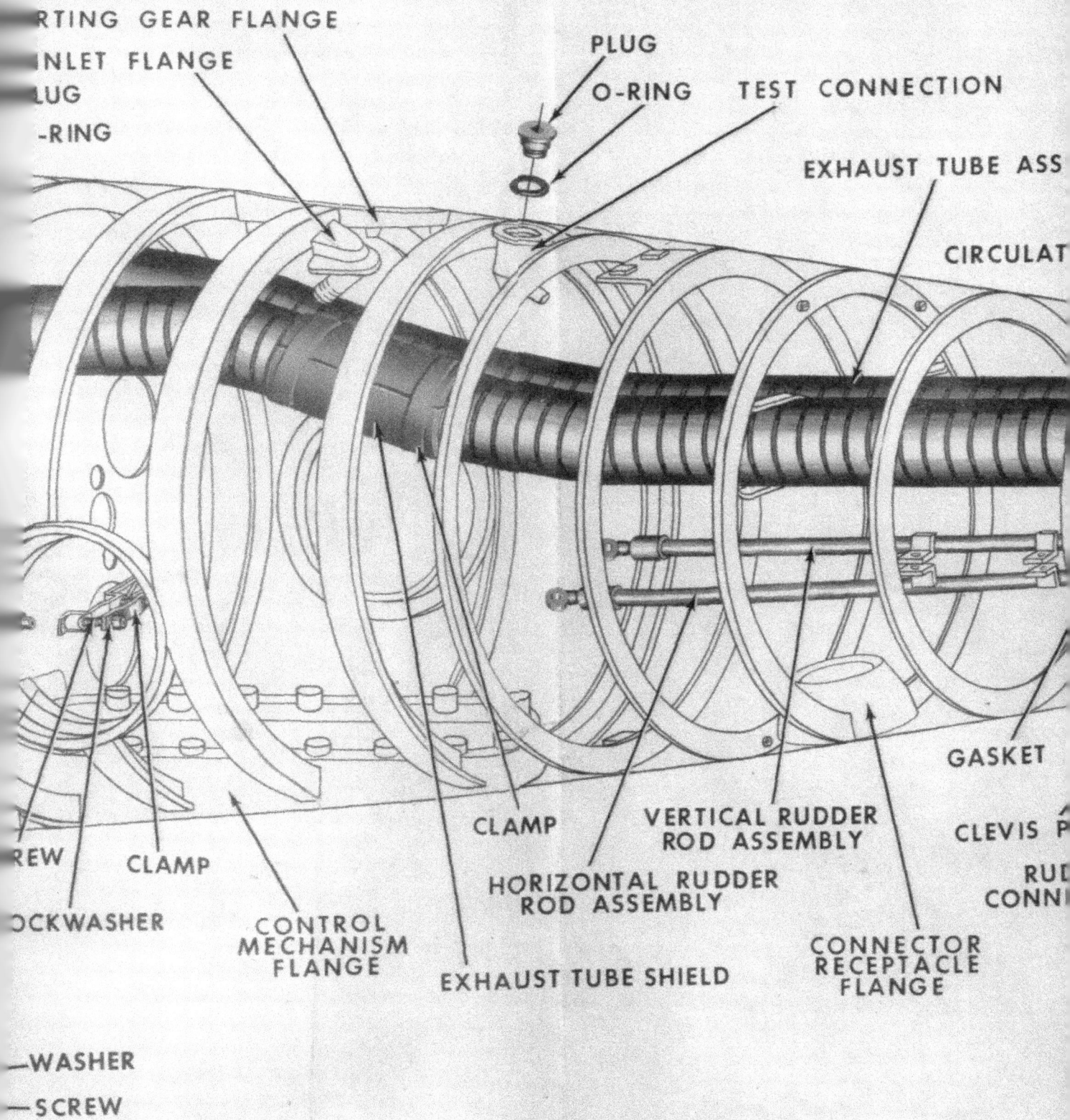

RTING GEAR FLANGE

NLET FLANGE

LUG

-RING

PLUG

O-RING   TEST CONNECTION

EXHAUST TUBE ASS

CIRCULAT

GASKET

CLEVIS P

RUD
CONN

REW   CLAMP

OCKWASHER

CONTROL
MECHANISM
FLANGE

CLAMP

HORIZONTAL RUDDER
ROD ASSEMBLY

VERTICAL RUDDER
ROD ASSEMBLY

CONNECTOR
RECEPTACLE
FLANGE

EXHAUST TUBE SHIELD

WASHER

SCREW

CONNECTION

EXHAUST TUBE ASSEMBLY

CIRCULATING PLATE

SCREW

AFTER
BULKHEAD
BEARING

EXHAUST VALVE ASSEMBLY

O-RING

GASKET

SCREW

RETAINER

O-RING

AFTER BULKHEAD

COTTER PIN

GASKET

CLEVIS PIN

RUDDER
CONNECTION

INSERT

O - RING

RETAINING SCREW

RUDDER
CONNECTION
END

SCREW

DDER
MBLY

CONNECTOR
RECEPTACLE
FLANGE

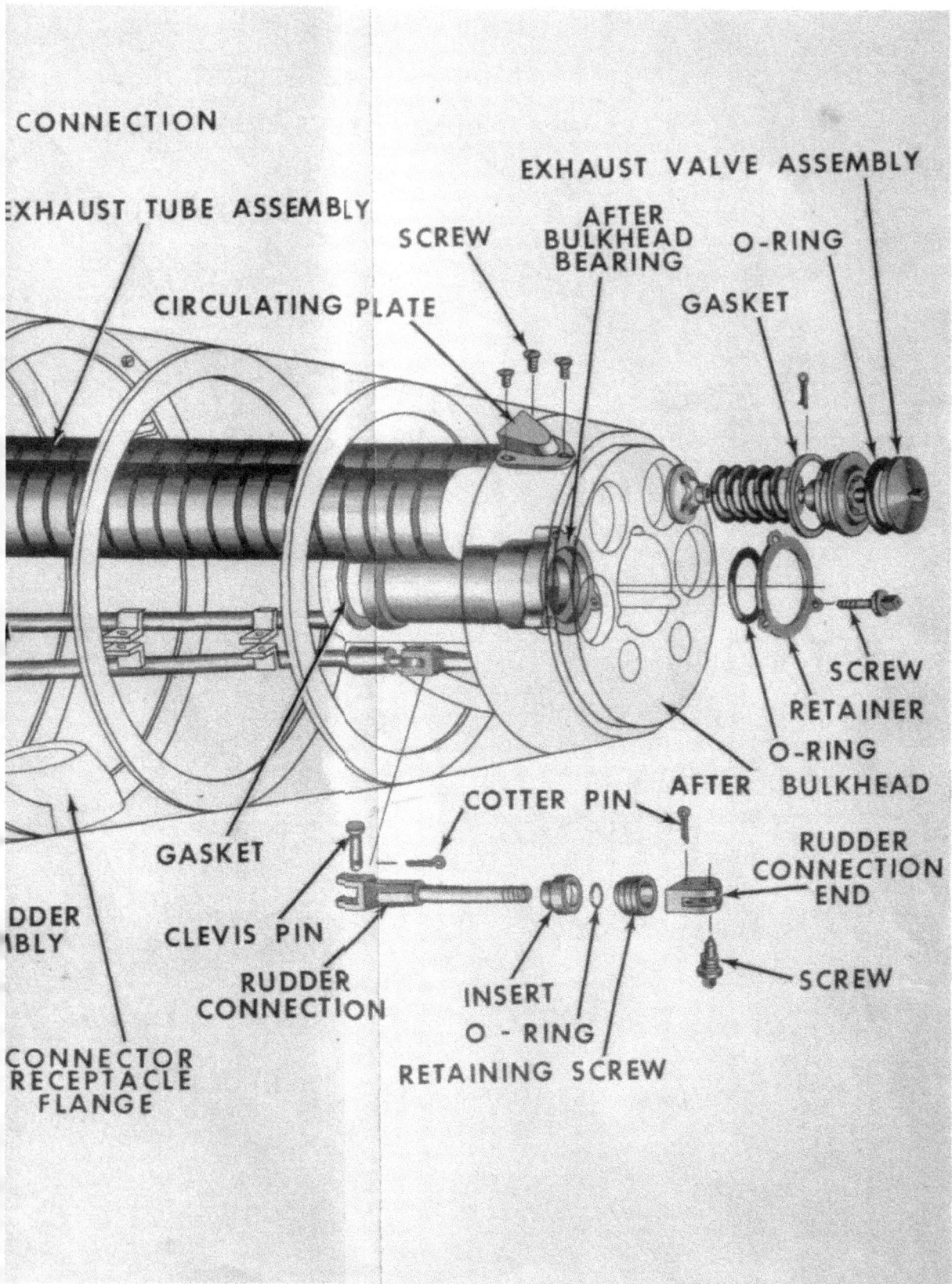

Figure 2-34. Afterbody Shell Assembly.

2-90. Vertical Bulkhead. The vertical bulkhead is a steel disk spotwelded to the shell internal surface at approximately one-third of the afterbody length from the forward end. It contains holes for installing the exhaust system, passage of pipe assemblies, and mounting the idler gear bracket, and it has two bushings for supporting the afterend of the turbine and gear train assembly.

2-91. Exhaust System. The exhaust system consists of a manifold assembly, two tube assemblies, and four valve assemblies. It carries the exhaust gases from the main engine turbines to the sea.

2-92. The manifold assembly is made of sheet steel pieces formed with an elliptical inlet at the forward end and two tubular outlets at the after end. A bracket, welded to the top of the manifold, secures the manifold to the forwardmost strengthening ring. Each tubular outlet fits into a tube in the vertical bulkhead.

2-93. The two tube assemblies are made of several cylindrical pieces of sheet steel interlocked to make the tubes flexible. A flange is welded to one end of each tube which is inserted through a hole in the vertical bulkhead. Each tube is also secured to the third strengthening ring from the afterend of the after body with a bracket and is inserted into flanged holes of the after bulkhead assembly.

2-94. Each of the four exhaust valve assemblies is a FIR unit and consists of a normally closed, spring-loaded valve held in place on a valve seat by a spring retainer. All parts are made of Monel. Each of the four valve seats is threaded into a hole in the after bulkhead assembly and sealed in place with a gasket. An O-ring is installed on the valve seat to insure a watertight seal between the mating faces of the valve and valve seat when the torpedo is in a flooded tube and when floating at the end of an exercise run. After the torpedo is fired, the O-ring becomes expendable. The exhaust valve assemblies open when the exhaust gases create a pressure of approximately 4 psi.

2-95. Handhole Covers. There are two circular handhole covers made of cast aluminum, contoured to conform with the outer surface of the afterbody shell, and equipped with an O-ring. When they are installed in the access handhole flanges of the afterbody shell, each one is secured in place with two screws and lockwashers, and its O-ring creates a watertight seal.

2-96. Test Connection. The test connection is a cylindrical flange welded to the top center of the afterbody shell. It permits testing of the pneumatic system in the afterbody. When not in use, it is sealed with an O-ring and a threaded steel plug.

2-97. After Bulkhead Assembly. The after bulkhead assembly consists of a welded steel body, a bronze after bulkhead bearing, and two stainless steel tubes. The circumference of the body is tapered to fit inside the afterend of the afterbody shell and is welded in place. When all relative components are installed in the afterbody bulkhead, the afterend of the afterbody bulkhead becomes watertight.

2-98. The body contains several holes, flanged, threaded, and counterbored as required, to accommodate the exhaust tubes, exhaust valves, after bulkhead bearing, rudder rod connections, and two tubes of a water cooling system. Two upper holes for the exhaust tubes pass into an open chamber in the bulkhead. The four exhaust valves are installed in four holes in the aft face of the after bulkhead and enter the open chamber. A central hole receives the after bulkhead bearing, a gasket and O-ring for watertight sealing, and a retaining ring and three screws for securing the bulkhead bearing in place. Two lower holes pass through the body to receive the rudder connections installed with an insert, O-ring, retaining screw, and end connection. A shoulder around the afterend perimeter of the body contains several threaded holes equally spaced and set at an angle to receive the studs that secure the tail assembly to the after bulkhead.

2-99. The water cooling system (figure 2-35) is composed of the holes and cavities contained in the two circulating plates (scoops), attached to the afterend of the afterbody shell, the two after bulkhead tubes, the channel around the exterior of the after bulkhead bearing, and the internal area of the tail assembly. When the torpedo is traveling, the scoops pick up sea water and the tubes circulate it around the bearing channel. Circulation is continuous because the sea water passes through openings in the after bulkhead body and through the tail assembly where it joins with exhaust gases before returning to the sea.

2-100. Rudder Rod Assemblies. There are two similar rudder rod assemblies, one horizontal and the other vertical, interconnecting the steering engines of the control mechanism with the rudders of the tail assembly. Each one consists of a straight rod with an eye connection on each of the forward and after ends. The forward eye connection is attached to the linkage of the steering engine. The after eye connection is attached to the rudder connection in the after bulkhead assembly. All components of the horizontal and vertical rudder rod assemblies are made of forged steel. However, the vertical rudder rod is made of steel with a very low coefficient of expansion.

2-101. COMBUSTION SYSTEM. The combustion system (figure 2-36) is mounted on the forward surface of the turbine bulkhead of the main engine (figure 2-33) and consists of the following major components:

1. Decomposition Chamber.

2. Combustion Pot.

3. Nozzle, Pipe, and Combustion Pot Bottom.

4. Two Igniters Mk 6 Mod 4.

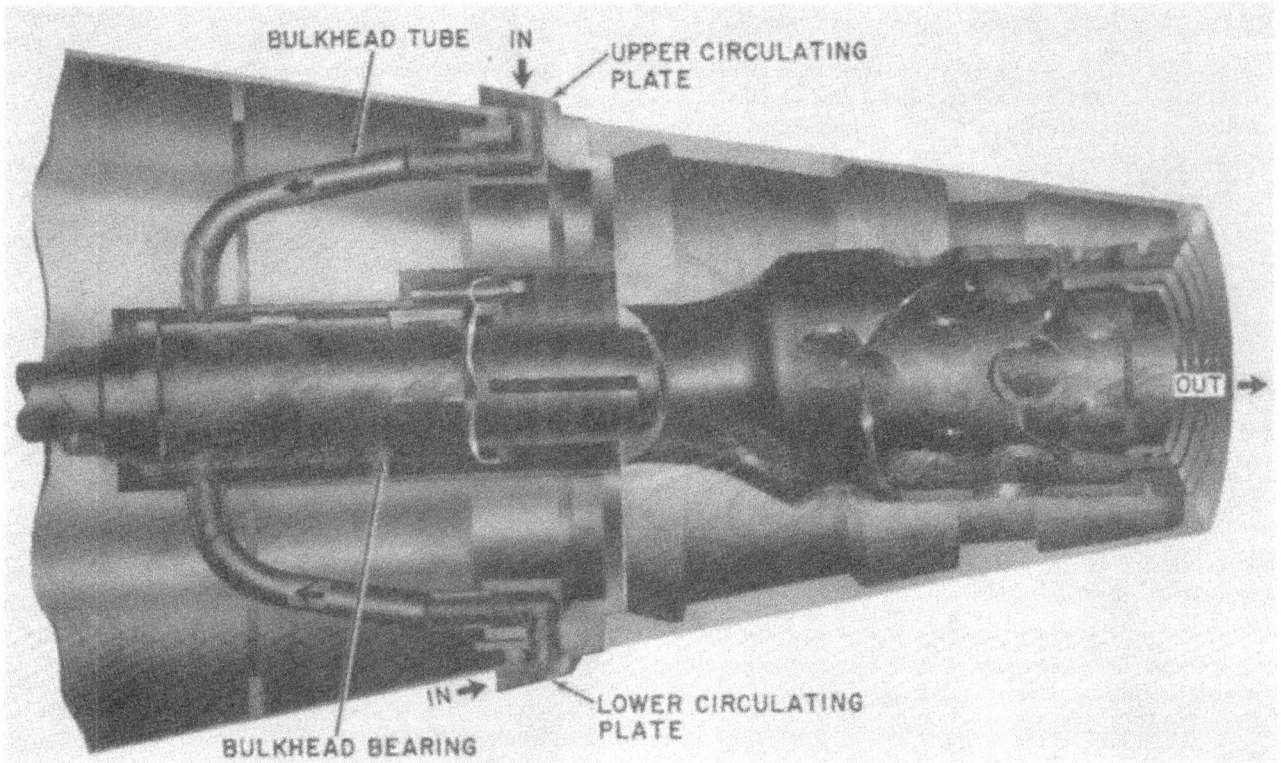

Figure 2-35. Water Cooling System of After Bulkhead.

2-102. Decomposition Chamber. The decomposition chamber (figure 2-36) is composed of the following parts:

    1. Body.

    2. Spacer.

    3. Catalyst Cartridge.

    4. Cover Assembly.

    5. Navol Restriction Assembly.

    6. Water Spray Assembly.

    7. Hollow Stud.

2-103. Body. The decomposition chamber body is made of a corrosive-resistant, cylindrical steel casting, completely open on one end and partially opened through a boss at the other end. Its internal cylindrical surfaces receive the spacer and catalyst cartridge. Its completely opened end receives the cover assembly. Its bossed end receives the water spray assembly and mates with a boss on the combustion pot body.

2-104. Spacer. The spacer is a cylindrical, corrosive-resistant steel sleeve inserted into the decomposition chamber body: to form a space for the accumulation of hot oxygen and steam, and to provide a shoulder for keeping the catalyst cartridge in its proper location.

2-105. Catalyst Cartridge. The catalyst cartridge is made up of approximately 100 catalyst screens arranged in three groups, separated by two piston rings and spider washers. The catalyst screens, piston rings, and spider washers are held together by two end plates and a bolt that passes through a center hole in the catalyst screens, spider washers, and end plates. Several holes are drilled through both end plates to allow entrance of Navol and exit of hot oxygen and steam. All the cartridge components, except the catalyst screens, are made of corrosive-resistant steel. The catalyst screens are made of a cobalt or silver material to act as the agent for decomposing the Navol.

2-106. Cover Assembly. The cover assembly consists of a cover, plug, O-ring, and washer. The cover is a circular corrosive-resistant steel casting with a protruding central boss and a nipple attached to the side of the boss. A threaded and machined hole through the center of the boss receives the washer and corrosive-resistant steel plug. The nipple has a machined hole passing through its center and an external thread used for the installation of the Navol restriction assembly. The larger face of the cover receives the O-ring, and around its perimeter are 12 counterbored holes to receive the screws that secure the cover to the decomposition chamber body.

2-107. Navol Restriction Assembly. The Navol restriction assembly regulates Navol delivery. It consists of the Navol restriction, which is a FIR part, and a restriction holder, both made of corrosive-resistant steel, a washer, and an O-ring. The Navol restriction is a small cylindrical nozzle with a shoulder, O-ring groove, and O-ring for its installation, and a small, tapered hole through its center for Navol delivery. The restriction holder has an external hexagon shape to accept a wrench, a machined internal surface to receive the washer, an internal thread to secure it to the nipple on the cover assembly, and an external thread to accept a pipe assembly.

2-108. Water Spray Assembly. The water spray assembly (figure 2-37) is a FIR part: it consists of the spray body, water restriction, O-ring, and spray extension; it regulates water delivery, and is installed in the hollow stud. Its metal parts are made of corrosive-resistant steel. The spray body has an external hexagon shape to accept a wrench, an external thread at one end to accept a pipe assembly, an external thread at the other end to secure it in the hollow stud, and a hole through its center, which is counterbored to receive the water restriction and is threaded to receive the spray extension. The water restriction is a small cylindrical nozzle with a shoulder, O-ring groove, and O-ring for its installation, and a small, tapered hole through its center for water delivery. The spray extension has a hexagonal shape to accept a wrench, an external thread at one end for installation, and a tubular extension at the other end to direct the water spray into the inlet of the combustion pot body.

2-109. Hollow Stud. The hollow stud serves three purposes: (1) to house the water spray assembly, (2) to interconnect the outlet of the decomposition chamber body and the inlet of the combustion pot body, and (3) to conduct hot oxygen and steam from the decomposition chamber body to the combustion pot body. It is made of corrosion-resistant steel and has a hexagonal shape to accept a wrench, an internal thread at one end to accept the water spray body, an external thread at the other end to secure the outlet of the decomposition chamber body to the inlet of the combustion pot body, and an internal chamber with six outlet holes on its circumference to form the conduit portion.

2-110. Combustion Pot. The combustion pot (figure 2-36) is composed of the following parts:

1. Body.

2. Fuel Spray Assembly.

3. Outer Liner and Inner Liner and Whirl Assembly.

2-111. Body. The combustion pot body is made of cast iron. It is a cylindrical shell completely open at one end and has four bosses, three of which are combined at the other end, and one of which is on top near the completely opened end. Also, a bracket is formed on the back side and is a part of the casting. Its internal cylindrical surface receives the outer liner. The face of the completely opened end accepts a washer and contains 12 tapped holes for screws that secure the combustion pot bottom to the body. Two of the three end bosses are internally threaded to receive the igniters. The other one of the three bosses is threaded to receive the fuel spray assembly and contains three holes to accept the screws that secure the inner and outer whirls in place. The single top boss is internally threaded to accept the hollow stud and, on its outer face, accepts a gasket upon the joining together of the decomposition chamber outlet boss to this inlet boss. The integrally cast bracket of the combustion pot body is attached to the decomposition chamber body by another bracket, similar to a link, and secured in place with lockwashers and screws.

2-112. Fuel Spray Assembly. The fuel spray assembly (figure 2-38) is a FIR part consisting of a fuel spray body, insert assembly, and spray nozzle assembly, all parts of which are made of corrosive-resistant steel; it delivers the fuel (alcohol) in a correct amount. The spray body has an external milled square shape to accept a wrench, an external thread at one end to accept a pipe assembly, an external thread at the other end to accept the spray nozzle, a centrally located external thread and shoulder to secure it in the central end boss of the combustion pot body, and a hole through its center, which is counterbored to accept the insert. The insert is a small cylindrical restriction with a small hole through its center for fuel delivery. The cylindrical spray nozzle has an internal thread at one end to secure it to the body, an externally milled square at the other end to accept a wrench, and a U-shaped piece of wire welded on the outlet end to atomize the fuel spray.

2-113. Outer Liner and Inner Liner and Whirl Assembly. The outer liner is a steel alloy tube externally machined to fit inside the combustion pot body and internally machined to accept the inner liner and whirl. When the outer liner is installed in the combustion pot body, a helical channel with approximately 2-1/2 loops is formed. The inner liner and whirl is a welded assembly fabricated from a steel alloy tube with longitudinal slots, a piece of flat steel alloy helix with approximately seven coils, and a cone of sheet steel alloy. When

LEGEND FOR FIGURE 2-36

1 WATER SPRAY ASSEMBLY
2 GASKET
3 STUD
4 GASKET
5 SPACER
6 DECOMPOSITION CHAMBER
7 CATALYST CARTRIDGE
8 O-RING
9 O-RING
10 PLUG
11 DECOMPOSITION CHAMBER ASSEMBLY
12 WASHER
13 SOCKET HEAD CAPSCREWS
14 NAVOL RESTRICTION
15 O-RING
16 RESTRICTION HOLDER
17 COVER ASSEMBLY
18 PISTON RING
19 SOCKET HEAD CAPSCREW

20 BRACKET
21 WASHER
22 NOZZLE, PIPE, AND COMBUSTION POT BOTTOM ASSEMBLY
23 COMBUSTION SYSTEM
24 TORPEDO IGNITER MK 6 MOD 4
25 WASHER
26 GASKET
27 SOCKET HEAD CAPSCREWS
28 WASHER
29 INNER LINER AND WHIRL ASSEMBLY
30 OUTER LINER
31 COMBUSTION POT BODY
32 COMBUSTION POT BODY ASSEMBLY
33 SOCKET HEAD CAPSCREWS
34 FUEL SPRAY ASSEMBLY
35 WASHER
36 LOCKWASHER

the inner liner and whirl assembly is installed in the outer liner, another helical channel is formed. An efficient combustion system exists because the hot oxygen, steam, and water spray (eventually, additional steam) become mixed quickly and thoroughly as they pass through the outer and inner helical channels and elongated slots to join with the ignited fuel.

2-114. Nozzle, Pipe, and Combustion Pot Bottom. The nozzle is made of a steel alloy forging with a combined elliptical and arc shape. It has several bosses machined to provide a threaded nipple for sensing the pot pressure, threaded openings for removal plugs used in cleaning the nozzle ports, and counterbored inlet opening for one end of the pipe. Around the circumference, it has a shouldered face with several spot-faced holes used in installing the combustion system on the turbine bulkhead of the main engine, and four tapered outlet ports arranged obliquely for directing the high-velocity hot gases and steam at the proper angle to spin the first turbine of the main engine. The pipe is a curved piece of steel alloy tubing welded to the inlet opening of the nozzle and to the outlet opening of the combustion pot bottom. The hemispherical combustion pot bottom is made of cast iron with a protruding boss counterbored to accept one end of the pipe and a flanged flat face with 12 counterbored holes around its circumference for screws that secure the nozzle, pipe, and combustion pot bottom to the combustion pot body.

2-115. Igniter Mk 6 Mod 4. Igniter Mk 6 Mod 4 (figure 2-39) is a pistol and double-fuze type device. Two igniters minimize the possibility of ignition failure. The igniters are installed in the two top bosses cast integrally at one end of the combustion pot body and are interconnected by a Y-shaped fitting and associated piping. Each igniter consists

mainly of an ignition stock (body) which contains an igniter tube, three charges of ignition load, packing, two primer caps, two firing pins, a spring housing assembly, a diaphragm, four washers, and a plug, all of which are installed in the top opening of the body. A perforated disk, a metallic seal, and a securing screw are installed in the bottom opening of the body. When the igniter is in its shipping condition, a plastic plug is installed on the inlet nipple of the plug. Igniters are sealed in individual containers and shall remain sealed until ready for use.

2-116. TURBINE AND GEAR TRAIN ASSEMBLY. The turbine and gear train assembly (figure 2-40) is installed in the forward end of the afterbody and consists of the following major components:

1. Turbine Bulkhead Assembly.
2. Turbine Bulkhead Bearing Assembly.
3. Engine Frame.
4. First Turbine Wheel and Spindle Assembly.
5. Spindle Casing, Second Turbine Wheel, and Spindle Assembly.
6. Forward Idler Gear Assembly.
7. Countershaft and Gearing Assembly.
8. After Idler Gear Assembly.
9. Forward Propeller Shaft and Gear Assembly.
10. After Propeller Shaft and Gear Assembly.
11. Engine Casing Assembly.

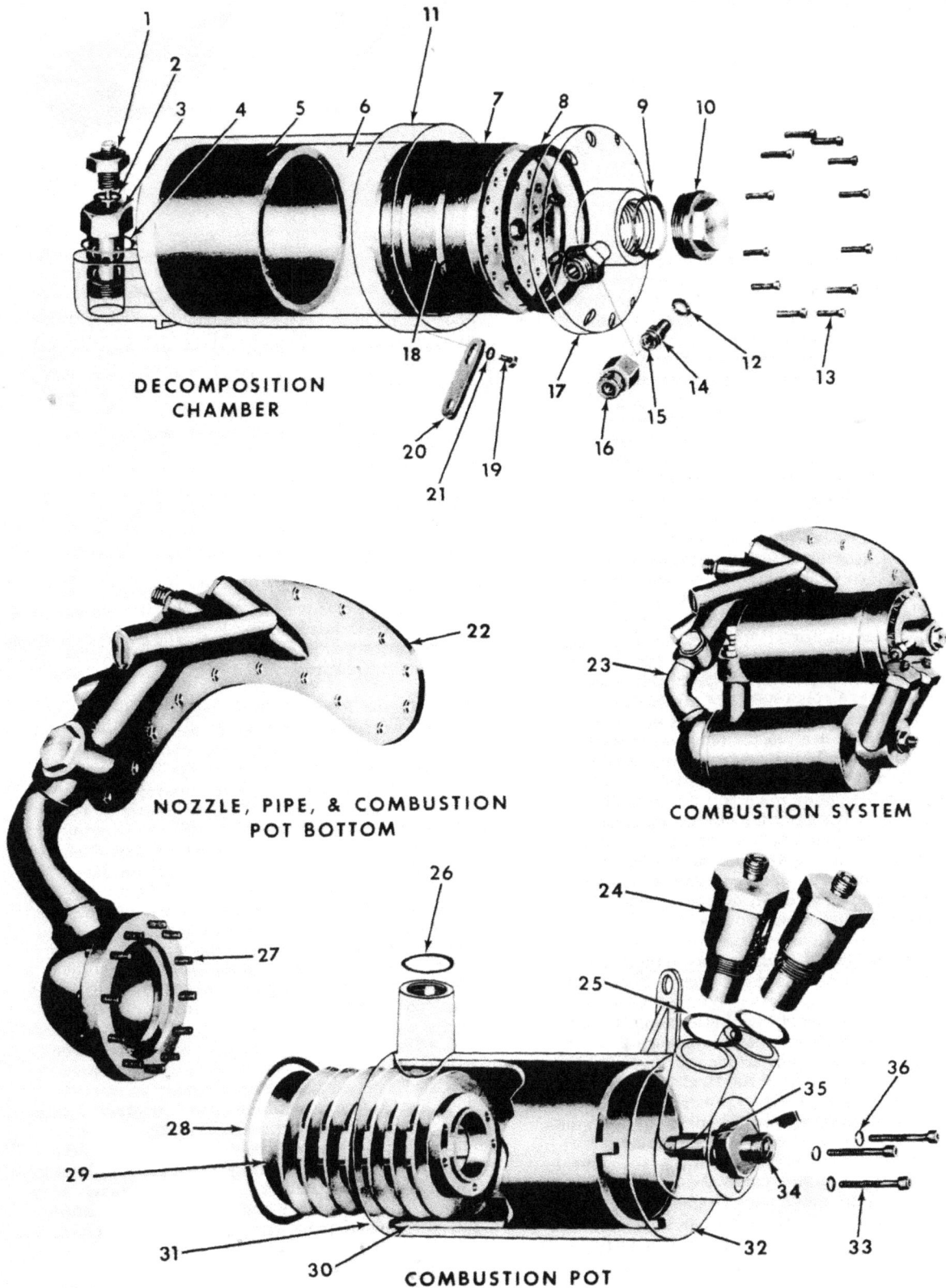

DECOMPOSITION CHAMBER

NOZZLE, PIPE, & COMBUSTION POT BOTTOM

COMBUSTION SYSTEM

COMBUSTION POT

Figure 2-36. Combustion System.

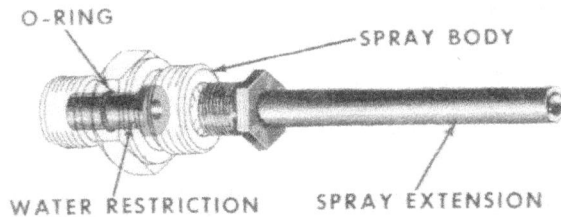

Figure 2-37. Water Spray Assembly.

Figure 2-38. Fuel Spray Assembly.

2-117. Turbine Bulkhead Assembly. The turbine bulkhead assembly consists of the turbine bulkhead and some of the parts and facilities necessary for the mounting of the combustion system, the turbine bulkhead bearing assembly, and the engine frame. When the turbine and gear train is completely assembled and installed, the turbine bulkhead assembly seals the forward end of the afterbody against entry of sea water. The turbine bulkhead is a circular cast steel plate with radial strengthening ribs, an elliptical arc-shaped flange, a center housing, and several bosses. The elliptical arc-shaped flange has machined surfaces and several studs used to secure the nozzle of the combustion system in place. The center housing has an outer machined face that contains several studs used for securing a gasket and cover in place and internal machined surfaces and threads for receiving the parts of the turbine bulkhead bearing assembly. There are two bosses, opposite each other, near the top of the bulkhead. One of these contains a low-pressure air pipe assembly and the other contains a nipple for installing a high-pressure air pipe assembly. One boss at one side and slightly below center of the bulkhead contains a tapped hole for the screw that secures the bracket on the combustion pot of the combustion system. Another boss at the center-lower perimeter of the bulkhead is machined and threaded to receive the gland fitting used to install the torpedo electrical cabling. The back face of the bulkhead is a completely machined flat surface with a shouldered perimeter for installing and sealing the bulkhead in the afterbody joint ring. Several tapped holes in the flat surface contain studs for securing the engine frame in place with locknuts.

2-118. Turbine Bulkhead Bearing Assembly. The turbine bulkhead bearing assembly (figure 2-41) consists of a ball bearing, a retainer, lockscrews, an alining sleeve, and a grease shell, all made of steel. It is installed in the housing at the center of the turbine bulkhead. It serves two purposes: (1) to guide and support the forward end of the first turbine wheel spindle, and (2) to adjust the clearance between the nozzle of the combustion system and the first turbine wheel. The ball bearing is a unit type consisting of balls retained by inner and outer races. The inner race fits snugly on the end of the first turbine wheel spindle, and the outer race fits snugly in the alining sleeve. Longitudinal position of the ball

bearing can be adjusted and secured by the retainer and lockscrews. External threads on the retainer and internal threads in the alining sleeve allow accurate positioning of the ball bearing. Slots in the retainer are compressed by the lockscrews to secure the threads after obtaining proper position of the ball bearing. The grease shield fits over the forward end of the first turbine wheel spindle and into the afterend of the alining sleeve. When the turbine bulkhead bearing assembly is completely installed, the gasket and cover on the forward end of the center housing of the turbine bulkhead and the grease shield retain the grease for lubricating the ball bearing. External threads on the afterend of the alining sleeve mate with threads inside the center housing of the turbine bulkhead. Turbine wheel-to-nozzle clearance is adjusted by screwing the alining sleeve in or out of the center housing. When a correct clearance adjustment is obtained, the alining sleeve is secured in place by two locking pins installed in the cover of the housing and inserted into notches at the forward end of the alining sleeve.

2-119. Engine Frame. The engine frame (figure 2-42) is a welded fabrication of ten main structural members, all made of steel, and consisting of a front plate, two front plate supports, two side plates, two top plates, and three bearing supports. Several rectangular and triangular pieces of steel are welded throughout the fabrication to provide structural strength. Each bearing support contains an upper and lower half-round cutout, machined integrally with a half-round cutout of its mating bearing cap. Studs in tapped holes of each bearing support the bearing caps, and locknuts installed on the studs secure the bearing caps in place. Most of the turbine and gear train components are installed on the bearing supports and their associated parts.

2-120. The front plate contains a central flanged hole to accommodate the forward end of the spindle casing and several tapped holes about its perimeter to install an oil retainer cover and gasket with securing screws and lockwashers. Around the perimeter of the top plate and secured in tapped holes are several studs to support a gasket and forward end of the engine casing secured in place with locknuts. The two front plate supports (left and right) are welded to the forward side of the front

The upper forward bearing surface and the forward portion of the upper center bearing surface support the spindle casing for the turbine wheel spindles. The after portion of the upper center bearing surface and the upper after bearing surface support the ball bearing shell and housing of the forward and after propeller shafts and attached gearing. The lower forward, center, and after bearing surfaces support the outer races of the ball bearings of the countershaft and gearing assembly. Two smaller bearing surfaces, one at the right of the upper forward bearing surface and the other at the right of the forward portion of the upper center bearing surface, support the forward idler gear shaft. Two other smaller bearing surfaces, one at the left of the lower center bearing surface and the other at left of the lower after bearing surface, support the after idler gear shaft. All the bearing surfaces insure proper meshing of the entire gear train.

2-122. Two tapped and counterbored holes in the afterend of the after bearing support accommodate two washers and studs for supporting the turbine and gear train at the vertical bulkhead in the afterbody. Threaded studs are secured in tapped holes at the afterend of the after bearing support for securing the afterend of the engine casing.

2-123. First Turbine Wheel and Spindle Assembly. The first turbine wheel and spindle assembly (figure 2-43) consists of the first turbine wheel, spindle, pinion gear, ball bearing, two locknuts, six keys, a lockwasher, and four lockscrews. The first turbine wheel is made of a stainless steel alloy capable of withstanding very high temperatures and contains several curved buckets around its outer circumference and a hub at its center. Several flat steel segments are riveted to the outer edges of the bucket and overlap to form a smooth outer circumference. The hub has a tapered center hole with keyways to install the turbine wheel on the spindle and a machined forward flat face with four drilled holes to accommodate the lockscrews that secure the turbine wheel locknut. The spindle is made of steel with a thread and bearing surface at its forward end to secure the spindle in the ball bearing of the turbine bulkhead bearing assembly, a thread and tapered portion with four keys set in keyways for securing the turbine wheel on the spindle. It has a central bearing surface to mate with the internal bearing surface of the second turbine wheel spindle, a shoulderd bearing surface with two keyways to accommodate the pinion gear and two keys, and a bearing surface and thread at its afterend to accommodate the ball bearing, the lockwasher, and the locknut. The pinion gear drives the countershaft and gearing assembly.

2-124. Spindle Casing, Second Turbine Wheel, and Spindle Assembly. The spindle casing, second turbine wheel, and spindle assembly (figure 2-44) consists of the spindle casing, the second turbine wheel, the spindle, an oil retainer, an oil ring retainer, two oil rings, two ball bearings, a spacer,

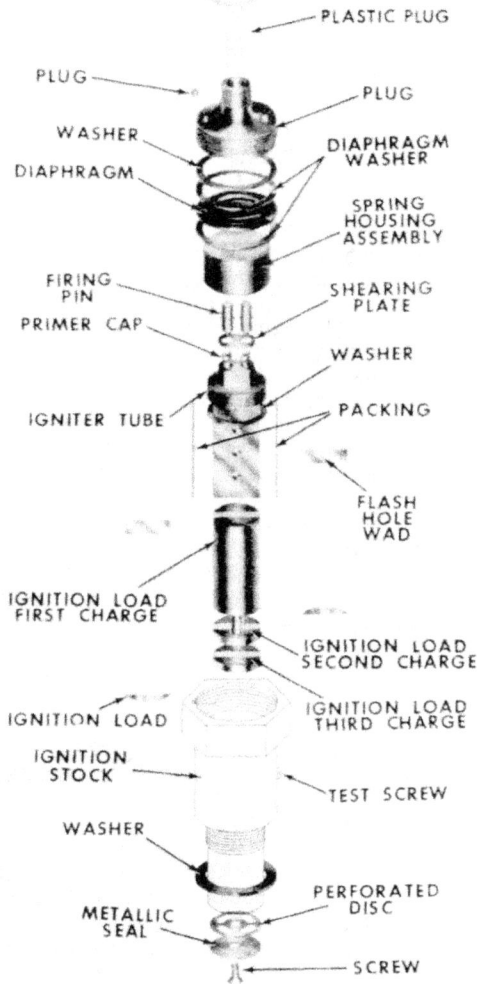

Figure 2-39. Igniter Mk 6 Mod 4.

plate. Each front plate support contains five holes to install the engine frame to the afterside of the turbine bulkhead. The two side plates (left and right) and two top plates (left and right) are welded to the afterside of the front plate and to each of the front, center, and after bearing supports to provide structural strength for the bearing supports. The forward bearing support is welded to the afterside of the front plate.

2-121. A complete bearing surface is formed when each bearing cap is installed on the studs adjacent to its designated upper or lower bearing support.

LEGEND FOR FIGURE 2-40

1 SELF-LOCKING NUT (1/2-20)
2 FORWARD BEARING CAP (UPPER)
3 CENTER BEARING CAP (UPPER)
4 AFTER BEARING CAP (UPPER)
5 BEARING SHELL
6 LOCKNUT (N-06)
7 LOCKWASHER (W-06)
8 NO. 306 BALL BEARING
9 AFTER PROPELLER SHAFT GEAR
10 BALL BEARING
11 HOUSING
12 PIN
13 AFTER PROPELLER SHAFT
14 KEY FOR AFTER PROPELLER SHAFT
15 LOCKNUT
16 LOCKWASHER
17 BUSHING
18 SPACER
19 FORWARD PROPELLER SHAFT GEAR
20 NO. 211 BALL BEARING
21 KEY FOR FORWARD PROPELLER SHAFT
22 FORWARD PROPELLER SHAFT
23 AFTER PROPELLER SHAFT BUSHING
24 PROPELLER SHAFT U-SEAL
25 U-SEAL RETAINER
26 SNAPRING
27 NUT
28 WASHER
29 AFTER PROPELLER SHAFT SCREWPLUG
30 IDLER GEAR SHAFT (FORWARD)
31 IDLER GEAR (FORWARD)
32 SPACER
33 TURBINE BULKHEAD ASSEMBLY
34 (UNASSIGNED)
35 (UNASSIGNED)
36 (UNASSIGNED)
37 LOCKNUT (1/2-28)
38 FORWARD COVER PLATE ASSEMBLY
39 GASKET FOR FORWARD COVER PLATE
40 LOCKNUT (N-05)
41 SCREW FOR TURBINE BULKHEAD RETAINER
42 RETAINER
43 LOCKWASHER (W-05)
44 NO. 405 BALL BEARING
45 ALINING SLEEVE
46 GREASE SHIELD
47 SCREW FOR TURBINE WHEEL NUT
48 TURBINE WHEEL LOCKNUT
49 FIRST TURBINE WHEEL ASSEMBLY
50 TURBINE WHEEL HUB KEY
51 FIRST TURBINE WHEEL SPINDLE ASSEMBLY
52 KEY FOR FIRST SPINDLE PINION
53 PINION - NO. 1 SPINDLE
54 SECOND TURBINE WHEEL ASSEMBLY
55 SCREW FOR OIL RETAINER
56 1/4 LOCKWASHER

57 OIL RETAINER
58 GASKET
59 OIL RING RETAINER
60 OIL RING
61 ADJUSTING RING FOR SPINDLE CASING FORWARD BEARING
62 LOCKSCREW
63 NO. 209 BALL BEARING
64 SPACER
65 SECOND TURBINE WHEEL SPINDLE
66 NO. 208 BALL BEARING
67 LOCKWASHER (W-08)
68 LOCKNUT (N-08)
69 ADJUSTMENT RING, AFT
70 LOCKSCREW
71 ENGINE FRAME
72 SPINDLE CASING
73 REINFORCING RING GASKET
74 GASKET FOR ENGINE CASING
75 ENGINE CASING ASSEMBLY
76 PIPE ASSEMBLY (BULKHEAD TO BRACKET)
77 UPPER CLAMP STRIP
78 FILLING AND DRAIN PLUG
79 REINFORCING RING FOR AFTEREND CASING
80 WASHER
81 SUPPORTING STUD
82 PIPE ASSEMBLY (BULKHEAD TO BRACKET)
83 LOCKNUT (1/40-20)
84 LOWER CLAMPING STRIP
85 LOCKWASHER SCREW UNIT
86 WHEEL WELL BAFFLE ASSEMBLY
87 LOCKNUT (N-07)
88 LOCKWASHER (W-07)
89 NO. 307 BALL BEARING
90 COUNTERSHAFT GEAR (FORWARD)
91 COUNTERSHAFT GEAR (AFTER)
92 NO. 308 BALL BEARING
93 COUNTERSHAFT SPACING WASHER (INNER)
94 COUNTERSHAFT SPACING WASHER (OUTER)
95 COUNTERSHAFT SPACING WASHER (AFTER)
96 COUNTERSHAFT
97 KEY FOR COUNTERSHAFT
98 SPACING COLLAR
99 NO. 207 BALL BEARING
100 SPACING COLLAR FOR AFTER IDLER GEAR
101 IDLER GEAR (AFTER)
102 IDLER GEAR SHAFT (AFTER)
103 FORWARD BEARING CAP (LOWER)
104 CENTER BEARING CAP (LOWER)
105 AFTER BEARING CAP (LOWER)
106 (UNASSIGNED)
107 (UNASSIGNED)
108 WASHER
109 GASKET

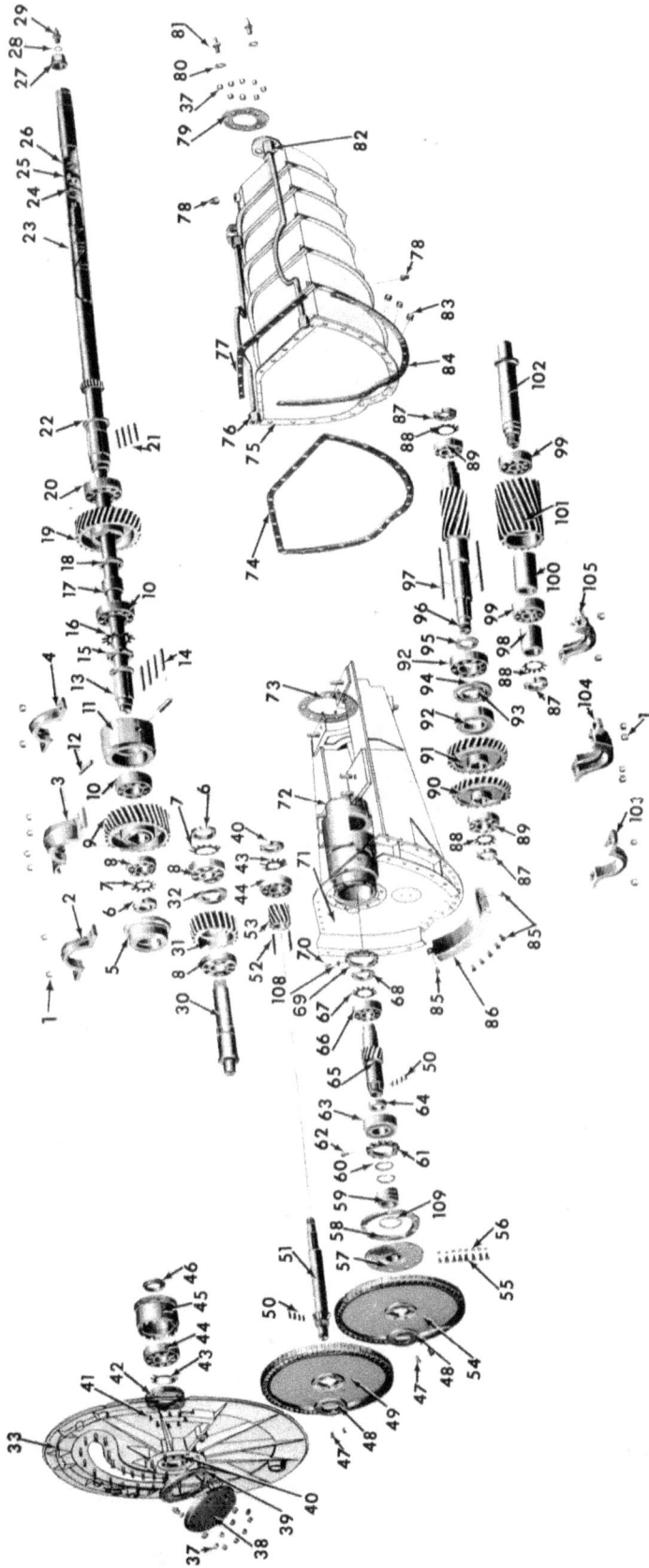

Figure 2–40. Turbine and Gear Train Assembly.

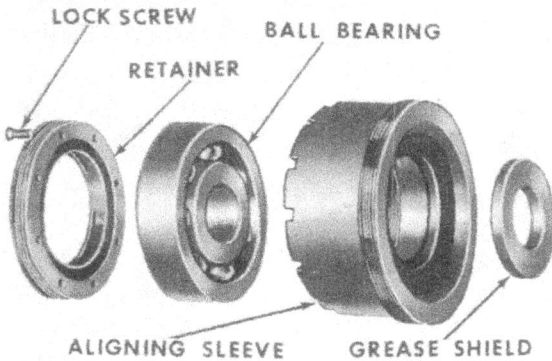

Figure 2-41. Turbine Bulkhead Bearing Assembly, Exploded View.

two adjusting rings, two locknuts, a lockwasher, four keys, and eight screws and lockwashers. The spindle casing is a cylindrical steel tube that houses all of the components assembled on the first and second turbine wheel spindles, except the turbine wheels and their locknuts. The spindle casing interior is counterbored to accommodate the outer races of the two ball bearings installed on the second turbine wheel spindle and the outer race of the ball bearing installed on the first turbine wheel spindle. A slot in the forward end of the casing fits over a dowel pin in the forward upper bearing support of the engine frame to insure accurate alinement. There are three cutouts in the central portion of the casing shell. The cutout near the forward end of the casing allows entry of the forward idler gear that meshes with the pinion gear on the second turbine wheel spindle. The cutout near the afterend of the casing allows entry of the pertinent countershaft gear that meshes with the pinion gear on the first turbine wheel spindle. The cutout near the center of the casing allows entry into the casing to set the adjusting ring used for positioning the forward ball bearing on the second turbine wheel spindle. At one end of the central cutout is a screw-type locking pin to lock the adjusting ring in place.

2-125. The second turbine wheel construction is similar to that of the first. However, its buckets are curved in the opposite direction to cause clockwise rotation (looking forward) and are deeper to provide extra surface for receiving the dispelled pressure of hot gases and steam from the buckets of the first turbine wheel. Although the buckets are deeper, the bottom of the buckets are on the same diameter for both turbine wheels. Consequently, the second turbine wheel has a larger outer diameter than the first turbine wheel.

2-126. The second turbine wheel spindle is made of steel and contains a hole through its center to accommodate the first turbine wheel spindle. An external thread and tapered bearing surface with four keyways are provided at the forward end of the spindle to mount four keys, the second turbine wheel, and the turbine wheel locknut. Four lockscrews pass through tapped holes in the locknut and insert into drilled holes in the outer face of the turbine wheel hub to secure the turbine wheel on the spindle. The pinion gear of the second turbine wheel is machined integrally at the center of the spindle. Two bearing surfaces, one forward and the other aft of the pinion gear, accommodate the inner races of the two ball bearings. The forward ball bearing is held in place by a spacer on the spindle and an adjusting ring threaded into the forward end of the casing. The after ball bearing is held in place by a lockwasher and a locknut installed on the threaded afterend of the spindle and another locking ring threaded into the afterend of the casing. Both adjusting rings position the outer races of the ball bearings to obtain the correct location of the second turbine wheel and spindle assembly necessary to provide clearance between both turbine wheels. Each adjusting ring has several slots equally spaced around its threaded perimeter to receive a lockscrew installed in tapped holes in the casing shell.

2-127. The oil retainer gasket and oil retainer plate are installed on studs at the outer face of the flanged hole in the front plate of the engine frame to prevent leakage of engine oil from the external area about the forward end of the spindle casing. The copperclad gasket, the oil ring retainer, and two oil rings are installed between the oil retainer plate on the engine frame and the forward ball bearing on the second turbine wheel spindle to prevent leakage of engine oil from the internal area about the forward end of the spindle casing.

2-128. Forward Idler Gear Assembly. The forward idler gear assembly (figure 2-45) consists of the forward idler gear, a shaft, two ball bearings, a spacer, a lockwasher, and a locknut, all made of a steel alloy. The forward idler gear is counterbored at both ends to accommodate the outer races of the two ball bearings. The spacer is installed between the two ball bearings to insure that the idler gear has free rotation. The solid shaft has a shoulder and bearing surface to accommodate the inner races of the two ball bearings and hole in the spacer. The inner races and spacer are secured to the shaft by the lockwasher and the locknut. External threads just aft of the shaft bearing surface accommodate the locknut. The forward idler gear assembly, installed in the engine frame, is located adjacent to the right of the spindle casing. Both ends of the shaft have smaller diameters than the remaining portions, and they are tightly secured in the smaller bearing holes provided in the upper forward and center bearing supports and mating bearing caps of the engine frame. The forward idler gear is driven in a counterclockwise direction (looking forward) by

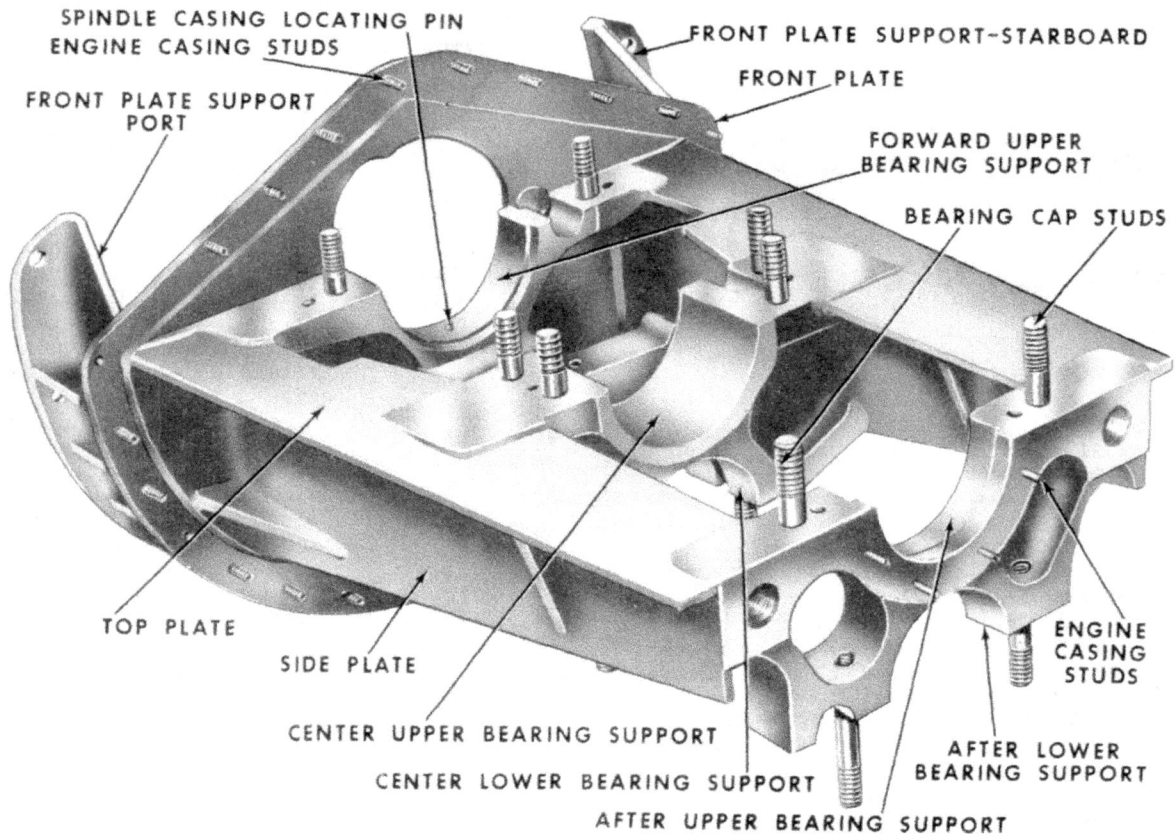

Figure 2-42. Engine Frame with Bearing Caps Removed.

the pinion gear of the second turbine wheel spindle. In conjunction with the pinion gear on the first turbine wheel spindle, it drives the countershaft and gearing in a clockwise direction (looking forward).

2-129. Countershaft and Gearing Assembly. The countershaft and gearing assembly (figure 2-46) consists of the countershaft, two gears, four ball bearings, three spacing washers, two keys, two lockwashers, and two locknuts, all made of a steel alloy. The countershaft is solid and has a pinion gear machined integrally with the remaining portions. The forward portion of the countershaft consists of four bearing surfaces that are progressively smaller in diameter with respect to their location from the pinion gear to the forward end of the countershaft. The after portion of the countershaft consists of two diametrical surfaces with the smaller one (a bearing surface) at the extreme afterend. The three spacing washers and the inner race of two ball bearings are installed on the bearing surface nearest to the forward end of the pinion gear. The next smaller bearing surface accommodates the two gears and two keys to lock the gears

on their axis. The next smaller bearing surface accommodates the inner race of the forwardmost ball bearing. The extreme forward bearing surface has an external thread and receives the lockwasher and locknut, which secure all the components installed on the bearing surfaces forward of the pinion gear. A portion of the countershaft just aft of the pinion gear does not contain a component but acts as a spacer. The extreme after bearing surface has an external thread and receives the inner race of the aftermost ball bearing and lockwasher and locknut which secure the inner race. The countershaft and gearing assembly is installed in the lower forward, center, and after bearing surfaces of the engine frame. The outer race of the forward ball bearing is secured by the lower forward bearing cap. The outer races of the two central ball bearings are secured by the lower center bearing cap. The outer race of the after ball bearing is secured by the lower after bearing cap. The countershaft and gearing assembly is driven in a clockwise direction (looking forward) by the combined torque of the first and second turbine wheels and through the pinion gear of the first turbine wheel spindle and the forward idler gear of the

Figure 2-43. First Turbine Wheel and
Spindle Assembly.

Figure 2-44. Spindle Casing, Second Turbine
Wheel, and Spindle Assembly.

second turbine wheel. The countershaft pinion gear drives the after propeller shaft counterclockwise, and the after idler gear drives the forward propeller shaft clockwise.

2-130. After Idler Gear Assembly. The after idler gear assembly (figure 2-47) consists of the after idler gear, a shaft, two ball bearings, two spacing collars, a lockwasher, and a locknut, all made of a steel alloy. The after idler gear contains a center hole to accommodate one of the spacing collars and is counterbored at both ends to accommodate the outer races of the two ball bearings. The solid shaft has a shoulder and central bearing surface to accommodate the inner race of the after ball bearing, one spacing collar, the inner race of the forward ball bearing, and the other spacing collar in their order of installation. The inner races and spacing collars are secured to the shaft by the lockwasher and locknut installed on external threads just forward of the central bearing surface of the shaft. The method of mounting the ball bearing allows the after idler gear to have free rotation about the shaft. The after idler gear assembly is installed in the engine frame and is located adjacent to and to the left of the after portion of the countershaft and gearing assembly. Both ends of the shaft have smaller diameters than the remaining portions and are tightly secured in the smaller bearing holes in the lower center, after bearing supports, and mating bearing caps of the engine frame. The after idler gear is driven in a counterclockwise direction (looking forward) by the pinion gear of the countershaft and gearing assembly. It drives the forward propeller shaft in a clockwise direction (looking forward).

2-131. Forward Propeller Shaft and Gear Assembly. The forward propeller shaft and gear assembly (figure 2-48) consists of the forward propeller shaft gear, two ball bearings, a spacer, a bushing, four keys, a lockwasher, and a locknut, all made of a steel alloy. The forward propeller shaft is hollow to accommodate the after propeller shaft. Near the forward end of the shaft is a pinion gear, which is an integrally machined part of the shaft, used to drive the idler gears mounted on a bracket attached to the vertical bulkhead in the afterbody. The extreme afterend of the shaft consists of a bearing surface with four keyways to accommodate the forward propeller sleeve housed in the tail assembly. Just forward of the afterend is a bearing surface that revolves in the bearing of the afterbody after bulkhead. Several holes through the bearing surface allow passage of lubricating grease. The forward end of the shaft has a shoulder, three bearing surfaces, and an external thread. The bearing surface nearest to the shoulder accommodates the inner race of the after positioned ball bearing. The central bearing surface accommodates the gear and has keyways that receive the four keys to lock the gear on its axis. The spacer is installed on the central bearing surface just forward of the gear. The remaining bearing surface accommodates the bushing on which is installed the

Figure 2-45.   Forward Idler Gear Assembly.

Figure 2-46.   Countershaft and Gearing Assembly.

Figure 2-47.   After Idler Gear Assembly.

inner race of the forward positioned ball bearing. The external threads at the forward end of the shaft accommodate the lockwasher and locknut, which secure in place the inner races, spacer, and gear. The outer races of the two ball bearings are secured to the engine frame, and the remainder of the forward propeller and gear assembly is free to rotate. The outer race of the forward ball bearing is installed in the floating ball-bearing housing attached to the after propeller shaft and gear assembly. The outer race of the after ball bearing is secured in the bearing surfaces of the upper after bearing support and mating bearing cap of the engine frame. The forward propeller shaft and gear assembly is driven in a clockwise direction (looking forward) by the after idler gear assembly. It rotates in the opposite direction of the after propeller shaft and gear assembly. Its after end is supported by the grease-retaining bushings installed between the afterends of the forward and after propeller and gear assemblies.

2-132.   After Propeller Shaft and Gear Assembly. The after propeller shaft and gear assembly (figure 2-49) consists of the solid after propeller shaft, a floating ball-bearing housing, a gear, two ball bearings, a bearing shell, a lockwasher, a locknut, a grease plug gasket, and a grease-retainer screw, all made of a steel alloy. Two bronze grease-retaining bushings, a U-seal, a U-seal retainer, and an internal snapring are not considered as parts of the after propeller shaft and gear assembly but are installed after the insertion of the after propeller shaft into the forward propeller shaft. At the extreme after end of the after propeller shaft is a bearing surface with four keyways to accommodate the after propeller sleeve housed in the tail assembly. A deep longitudinal hole in the extreme afterend of the shaft is counterbored and threaded to accommodate the grease plug and gasket. Just forward of the shaft afterend is a bearing surface that revolves in the grease-retaining bushings. Several holes pass through the bearing surface into the deep hole. The grease plug is removed when lubricating grease is applied, and the grease enters the deep longitudinal hole, passes through the holes in the bearing surface to lubricate the grease-retaining bushings and inner bearing surface of the forward propeller shaft. It then passes through holes in the bearing surface of the forward propeller shaft to lubricate the bearing in the afterbody after bulkhead. At the extreme forward end of the shaft is a shoulder, two bearing surfaces, and an external thread. The bearing surface nearest to the shoulder accommodates the gear and has four keys to lock the gear on its axis. At the afterend of the gear hub is a bearing surface to accommodate the tightly fitted inner race of the after ball bearing. The outer race of the after ball bearing is installed tightly in the forward end of the floating ball-bearing housing. The other bearing surface accommodates the inner race of the forward ball bearing. The bearing shell is installed on the outer race of the forward ball bearing. The lockwasher and locknut are installed on the external

Figure 2-48. Forward Propeller Shaft
and Gear Assembly.

Figure 2-49. After Propeller Shaft
and Gear Assembly.

threaded end to secure the inner race of the forward ball bearing and gear on the shaft. The outer race of the forward ball bearing and outer bearing surface of the floating ball-bearing housing are secured in the engine frame, and the remainder of the after propeller shaft and gear assembly is free to rotate. The outer race of the forward ball bearing is secured in the bearing surfaces of the upper center bearing support and mating bearing cap of the engine frame. The floating ball-bearing housing has two pins secured perpendicularly in tapped holes opposite to one another on the external housing surface. The pins rest on a portion of the engine frame to prevent rotation of the housing.

2-133. Engine Casing Assembly. The engine casing assembly (figure 2-50) is a ribbed, sheet-steel housing that encloses the gear train and retains oil for gear train lubrication. The casing is provided with accessories consisting of an upper clamp strip, a lower clamp strip, two nipple brackets, two threaded plugs, a reinforcing ring, two supporting studs, two stud washers, several locknuts, four gaskets, and two pipe assemblies. The engine casing gasket, the forward end of the engine casing, the upper clamp strip, and the lower clamp strip are secured to the studs on the after perimeter of the engine frame front plate with 26 locknuts. The circular flat gasket, the afterend of the engine casing, and the reinforcing ring are secured to the studs on the after bearing support of the engine frame with eight locknuts. The two supporting studs also secure the

afterend of the engine casing to the engine frame and support the afterend of the turbine and gear train at the afterbody vertical bulkhead. The two nipple brackets are secured to the upper afterend of the engine casing and accommodate the low- and high-pressure air pipe assemblies installed between the nipples on the brackets and nipples on the turbine bulkhead. Near the upper aft end of the engine casing is a flange containing a counterbored and threaded oil-filling hole. Near the lower afterend of the engine casing is another flange containing a counterbored and threaded oil-draining hole. When either the filling or draining hole is not in use, it is sealed with a gasket and threaded plug.

2-134. CONTROL MECHANISM. The control mechanism (figure 2-51) is made up of two groups of components. One group, called the gyro mechanism, presets and governs the torpedo running course. The other group, called the immersion mechanism, presets and governs the torpedo running depth. The gyro and immersion mechanisms (figures 2-52 through 2-55) are mounted on a common base, called the gyro pot and base assembly. As a unit, the control mechanism is installed in the lower flange of the afterbody shell (figure 2-33) and consists of the following major components:

   1. Gyro Pot and Base Assembly.

   2. Gyro Mechanism.

       a. Gyro Mk 12 Mod 3.

       b. Top Plate, Pallet Mechanism, and Bearing Assembly.

       c. Bottom Head, Clamp Plate, and Covers Assembly.

       d. Stop Block Assembly.

       e. Spinning and Unlocking Mechanism.

       f. Bracket and Pallet Driving Gear Assembly.

       g. Valve Rock Shaft Assembly.

       h. Gyro Reducing Valve.

       i. Vertical Steering (Course) Engine.

       j. Rudder Control Valve.

       k. Gyro Setter Assembly.

   3. Immersion Mechanism.

       a. Pendulum and Casing Assembly.

       b. Buffer Springs and Tension Rod Assembly.

       c. Depth Spring Assembly.

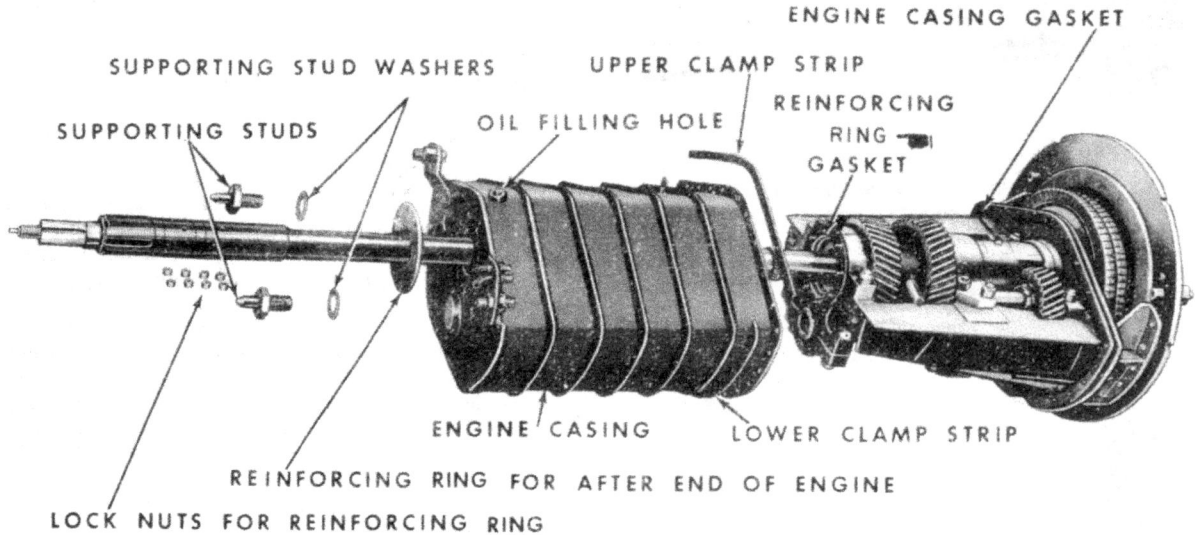

Figure 2-50. Engine Casing Assembly.

d. Air Chamber and Diaphragm Assembly.

e. Auxiliary Depth Spring.

f. Climb-Angle Limiter Device.

g. Transportation and Replacement Screws.

h. Horizontal Steering (Depth) Engine.

i. Depth Setter Assembly.

2-135. Gyro Pot and Base Assembly. The gyro pot and base assembly (figure 2-53) of the control mechanism is made up of an elongated, oval bronze casting to which is brazed a hollow bronze cylinder. The bottom of the base casting is curved and tapered to conform to the contour of the afterbody shell. Around the perimeter of the base casting are 18 counterbored holes to accommodate screws for securing a gasket and the control mechanism in the flange of the afterbody shell. At the lower forward end of the base casting is a cavity with a round flange having tapped holes to accommodate screws for securing the diaphragm ring of the air chamber and diaphragm assembly. At the upper forward end of the base casting is a round flanged hole with 11 studs to accommodate nuts for securing a gasket and the pendulum and casing assembly. The hollow cylindrical (gyro pot) portion houses Gyro Mk 12 Mod 3 and has several pads, bosses, and holes to allow installation of some components related to

the gyro. The top of the gyro pot is counterbored to receive the bearing of the top plate, pallet mechanism, and bearing assembly. The bottom of the gyro pot is counterbored and has an interrupted ridge to accommodate the bottom head, clamp plate, and covers assembly. At the after side of the gyro pot is a pad with six tapped holes to accommodate screws for securing the spinning and unlocking mechanism. At the forward side of the gyro pot is an oblong pad with three tapped holes to accommodate screws for securing the bracket and gearing assembly. At the port side (looking forward) of the gyro pot is a channeled boss with two tapped holes to accommodate screws for securing the horizontal steering engine. At the starboard side of the gyro pot is a similar channel boss with two tapped holes to accommodate screws for securing the vertical steering engine. At the upper inside port wall of the gyro pot is a curved pad with two holes to accommodate screws for securing the stop block assembly.

2-136. Gyro Mechanism. The gyro mechanism (figures 2-52, 2-53, and 2-54) consists of Gyro Mk 12 Mod 3 and affiliated components, some used alone and some in combination with others to provide gyro installation, gyro operation, presetting of gyro angle (course) order, and governing of course steering. Most of the gyro mechanism components are installed at the after area of gyro pot and base assembly. Remaining components are installed remotely because of design.

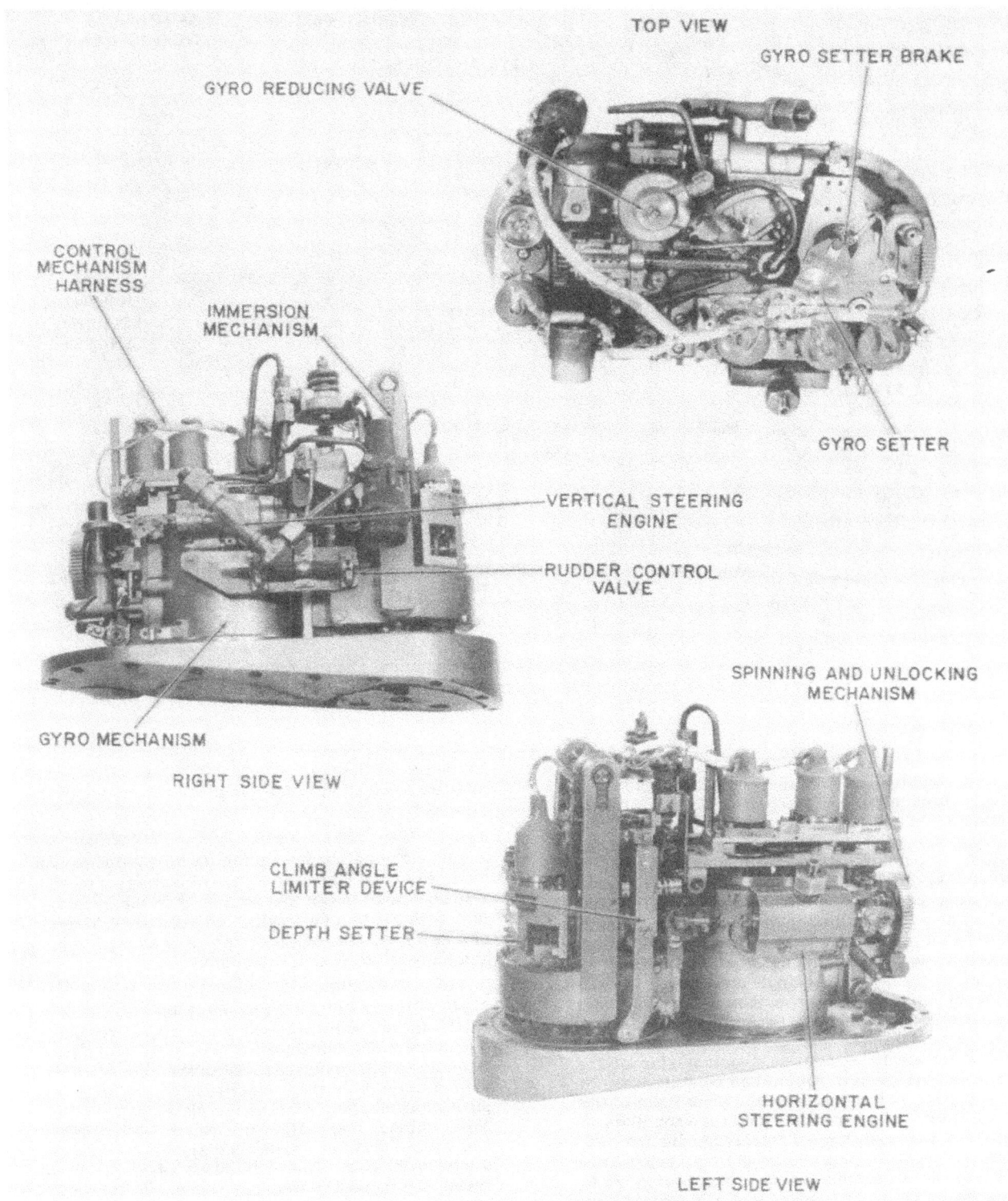

Figure 2-51. Control Mechanism.

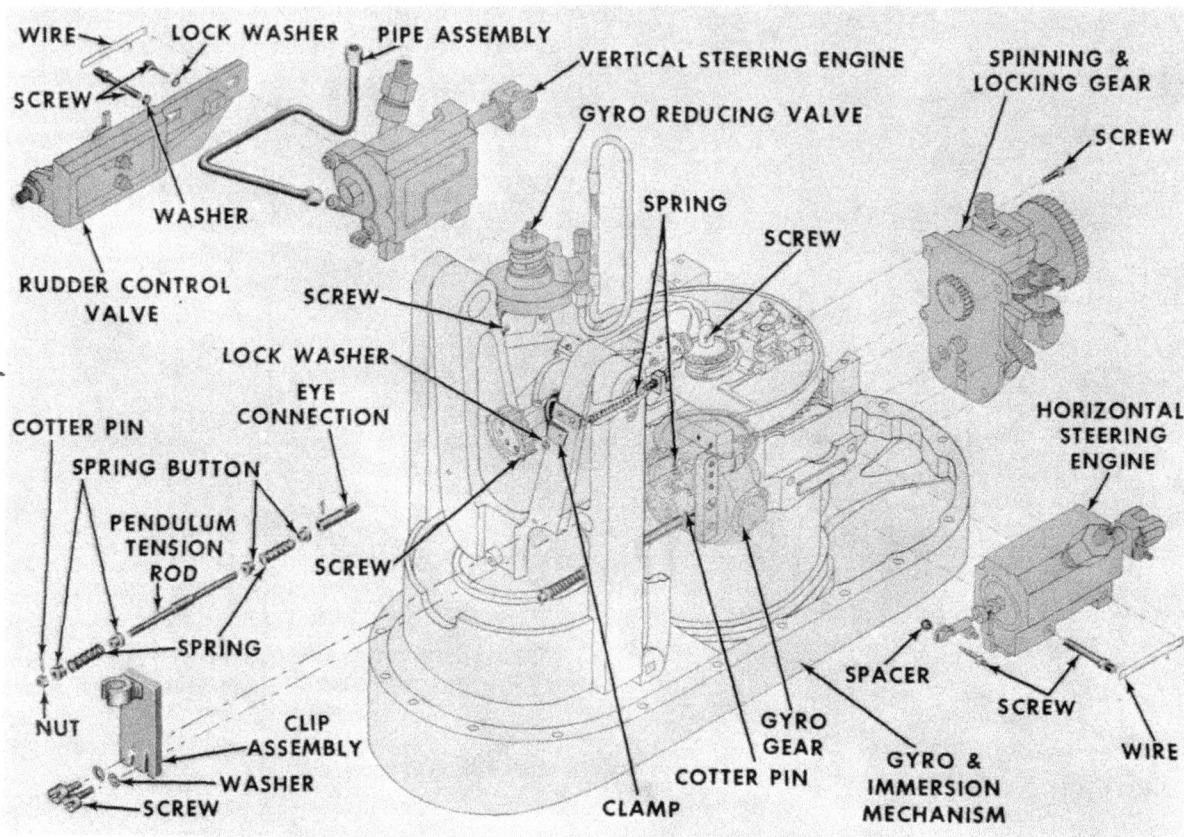

Figure 2-52. Control Mechanism, Mechanical Components.

2-137. Gyro Mk 12 Mod 3. The gyro (figure 2-56) is a nontumble type. Refer to OP 627 for a detailed physical description of the gyro.

2-138. Top Plate, Pallet Mechanism, and Bearing Assembly. The top plate, pallet mechanism, and bearing assembly (figures 2-53 and 2-57) is installed as a unit on the top of the gyro pot. The top plate has a top bearing holder installed in its bottom center to support the ball bearing on top of the outer gimbal of the gyro. A nipple and O-ring installed in the center hole of the top plate allows low-pressure air to pass through a center hole in the top bearing holder and enter the passage holes in the gyro components to provide constant spin of the gyro. A spur gear is cut on the circumference of the top plate and is driven by the gear train of the gyro setter. The upper portion of the top plate is recessed and contains a central spindle and two pads for the mounting of the pallet mechanism. The top plate bearing consists of an outer race secured to the bottom perimeter of the top plate with six screws and lockwashers and an inner race

and balls installed in the top inner perimeter of the gyro pot.

2-139. The pallet mechanism (figure 2-57) serves as an intermediate means for transmission of off-course signals from the cam plate on the gyro to the vertical steering engine. It consists mainly of a pallet shaft assembly, two pawl assemblies, and an oscillating drive assembly.

2-140. The pallet shaft assembly is made up of a shaft with two cam pawls on the shaft bottom end, a leaf spring on the shaft forward side, and a pallet on the shaft upper end. With the cam pawls at the bottom side of the top plate, the shaft passes through alining holes in the top plate, pallet slide, pallet holder, and pallet slide cover. The shaft leaf spring presses against the side of the hole in the pallet holder. The pallet shaft assembly is secured within its related components by the pallet, which is clamped to the shaft upper end by a screw and nut.

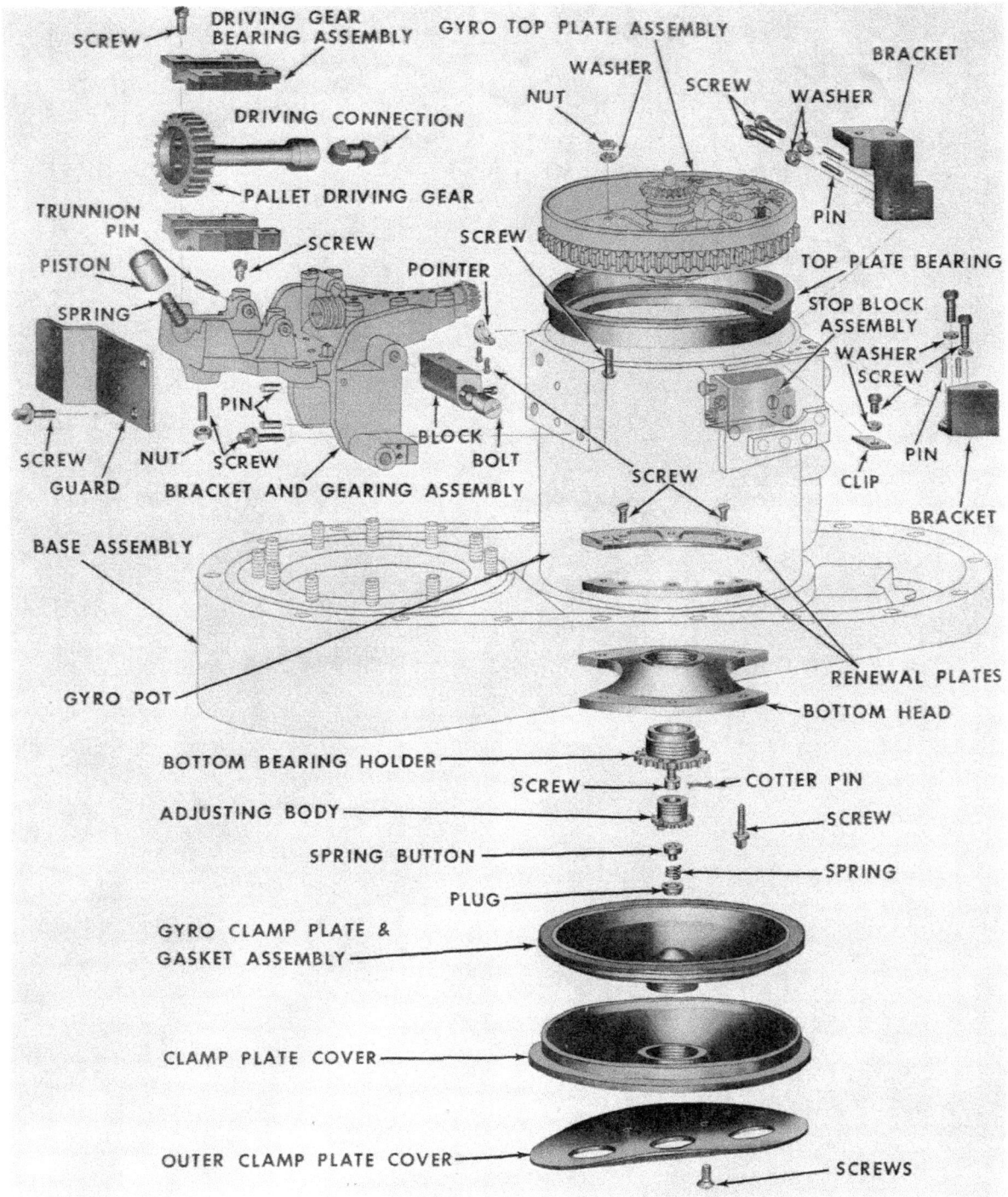

Figure 2-53. Gyro Mechanism, Partial Components.

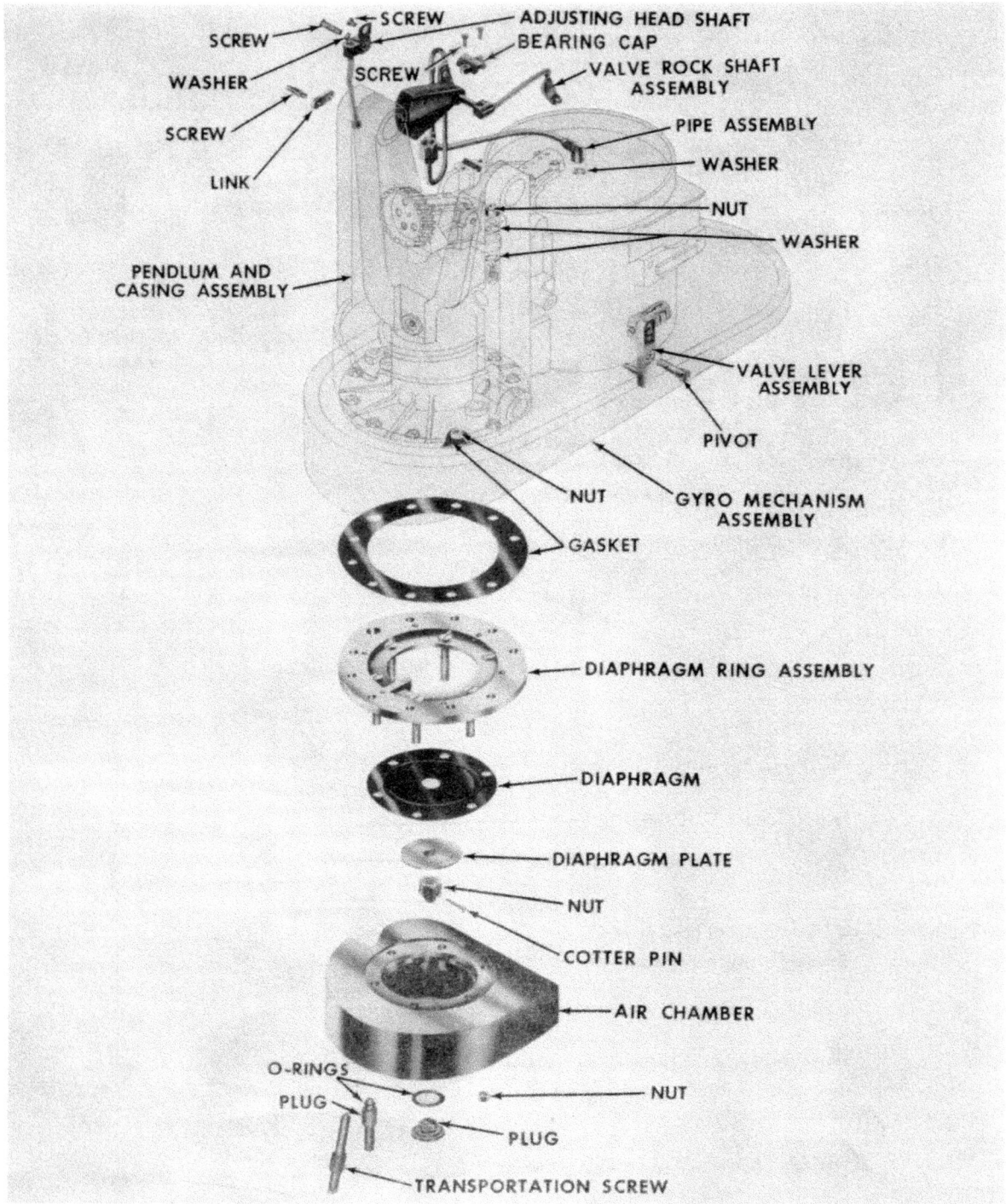

Figure 2-54. Gyro and Immersion Mechanism, Partial Components.

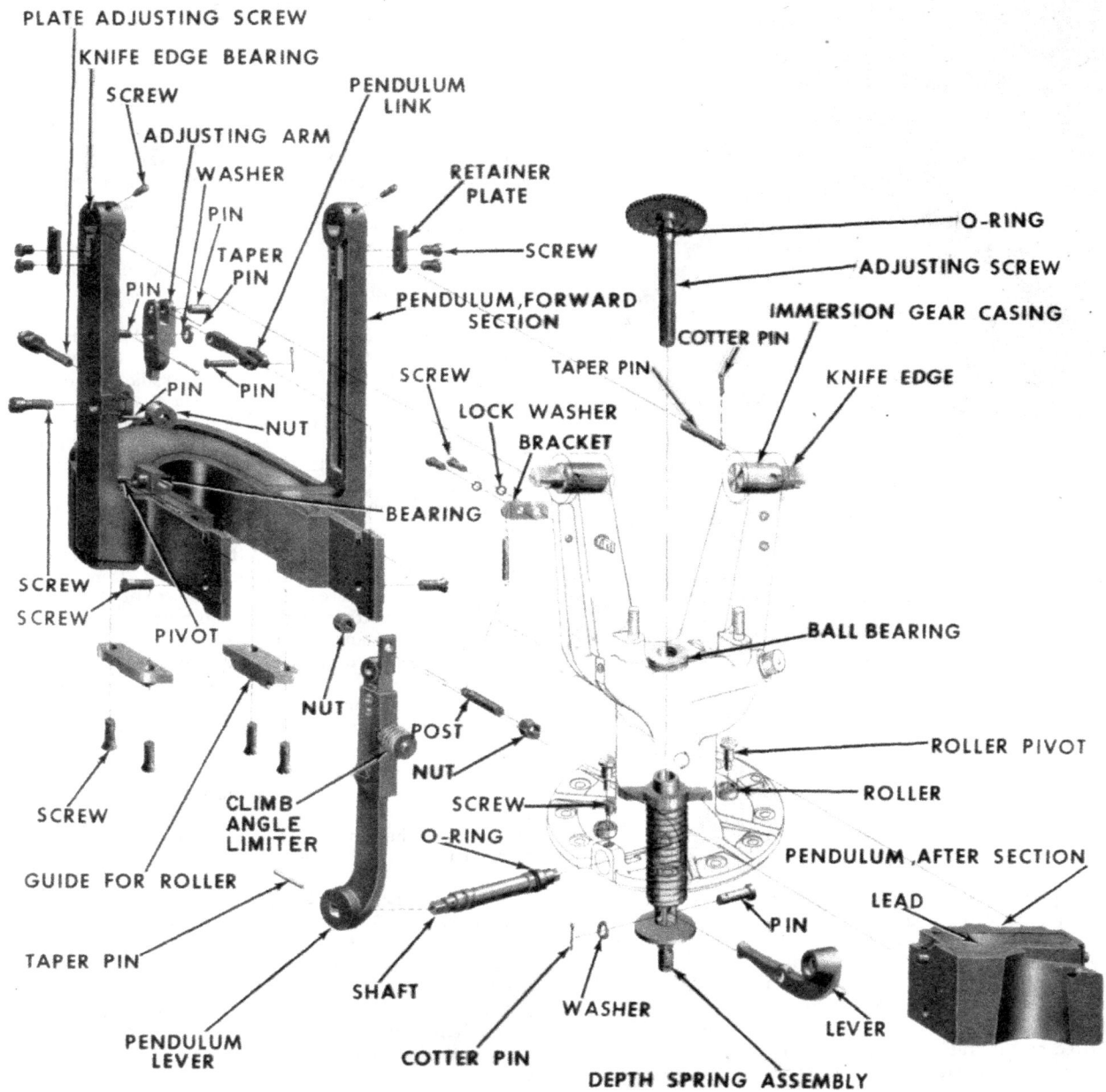

Figure 2-55. Pendulum and Casing Assembly, Exploded View.

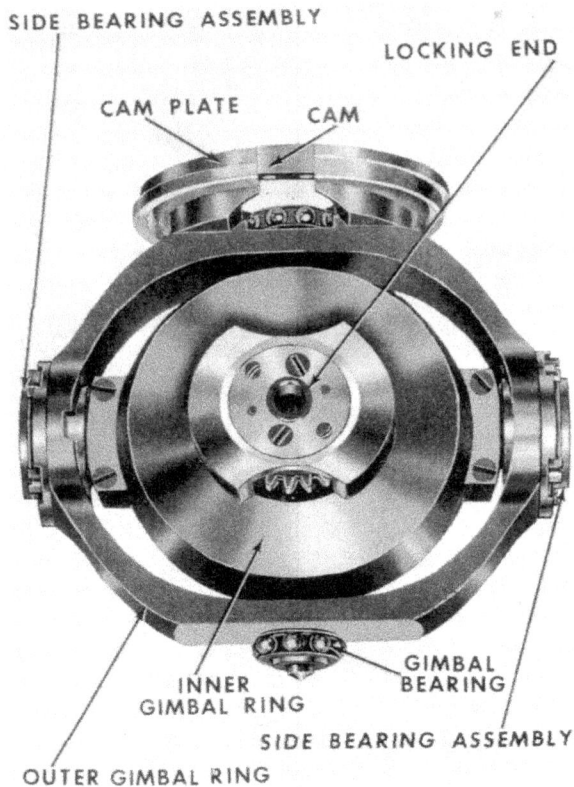

SIDE BEARING ASSEMBLY

LOCKING END

CAM PLATE   CAM

INNER GIMBAL RING

GIMBAL BEARING

SIDE BEARING ASSEMBLY

OUTER GIMBAL RING

Figure 2-56. Gyro Mk 12 Mod 3.

2-141. The two pawl assemblies are made up of a right-hand pawl, a left-hand pawl, two adjusting links, and an eye connection. The two pawls, each with a coil spring and washer, are installed on the two top posts of the pallet slide cover and are held in place by cotter pins. The two adjusting links are installed on their applicable upper forward posts of the two pawls, are held in place by cotter pins, and are interconnected by an adjusting screw. The eye connection has one end installed on the upper side post of the left-hand pawl, and the other end is installed on the side post of the pallet slide cover. Both eye connection ends are held in place with cotter pins. The length of the eye connection is adjustable by a built-in adjusting screw and locking arrangement.

2-142. The oscillating drive assembly is made up of a cam, a pallet slide, a pallet holder, a connection spool, a pallet slide cover, and a cam gear. The cam slides over the top center post of the top plate, rides in the elongated hole at the forward end of the pallet slide, is held in place by the connection spool and cam gear, and is driven by the cam gear. The afterend of the pallet slide is channeled and has a plain hole and a threaded hole to accommodate the installation of the pallet holder. The external sides

of the pallet slide channel ride in the space between the two pads at the upper afterend of the top plate. The connection spool has a center bearing hole to accommodate the cam gear shaft and a groove on its lower perimeter to accommodate the bell crank of the pallet slide cover and an upper groove for the valve rock shaft arm. The cam gear has a center bearing hole that rides on the upper center post of the top plate, a shaft with a bearing surface to accommodate the connection spool, and a groove in its bottom end to drive the cam. It has a bevel gear cut on its top end to mesh with the bevel drive gear of the bracket and gearing assembly. The nipple on the pipe assembly of the valve rock assembly secures the cam gear in place.

2-143. Bottom Head, Clamp Plate, and Covers Assembly. The bottom head, clamp plate, and covers assembly (figure 2-53) secures Gyro Mk 12 Mod 3 in the bottom of the gyro pot portion of the common base. The bottom head, together with two renewal plates, is secured across the bottom of the gyro pot by six screws. A bottom bearing holder and its associated parts are mounted in the center of the bottom head and support the lower ball bearing on the outer gimbal ring of the gyro. The clamp plate and covers assembly consists of the gyro clamp plate and gasket assembly, used to lock the gyro clamp plate in place, and the outer clamp-plate cover to conform with the bottom contour of the common base. The clamp plate and gasket assembly has a four-section interrupted ridge that engages a corresponding ridge in the bottom of the gyro pot. A threaded boss on the bottom of the clamp plate is secured into a central tapped hole in the clamp-plate cover. Thus, the four-section interrupted ridge permits removal and seating of the clamp plate by a 1/4-turn movement. When the clamp plate is properly seated, the tightening effect of the threaded boss and tapped hole forces the ridge seat of the clamp-plate cover against the mating ridge seat in the gyro pot.

2-144. Stop Block Assembly. The stop block assembly (figure 2-53) is fastened to the inside wall of the gyro pot and is adjacent to the horizontal steering engine mounting boss. Together with a T-shaped block secured to the underside of the gyro pot top plate, it prevents the top plate from turning more than 155° left or right of zero setting. Two spring-loaded sleeves in the stop block assembly cushion the impact when the top plate reaches its limit of travel.

2-145. Spinning and Unlocking Mechanism. The spinning and unlocking mechanism (figure 2-58) is mounted on the after side of the gyro pot (figure 2-52). It consists of the following principal parts: (1) gyro centering device, (2) spinning wheel, gear, and impulse mechanism, (3) locking and unlocking gear train, and (4) duration of spin (unlocking time) adjustment. The principal parts are mounted on a frame and front plate.

2-146. The gyro centering device locks the gyro in a fixed position (figure 2-59) by use of its centering

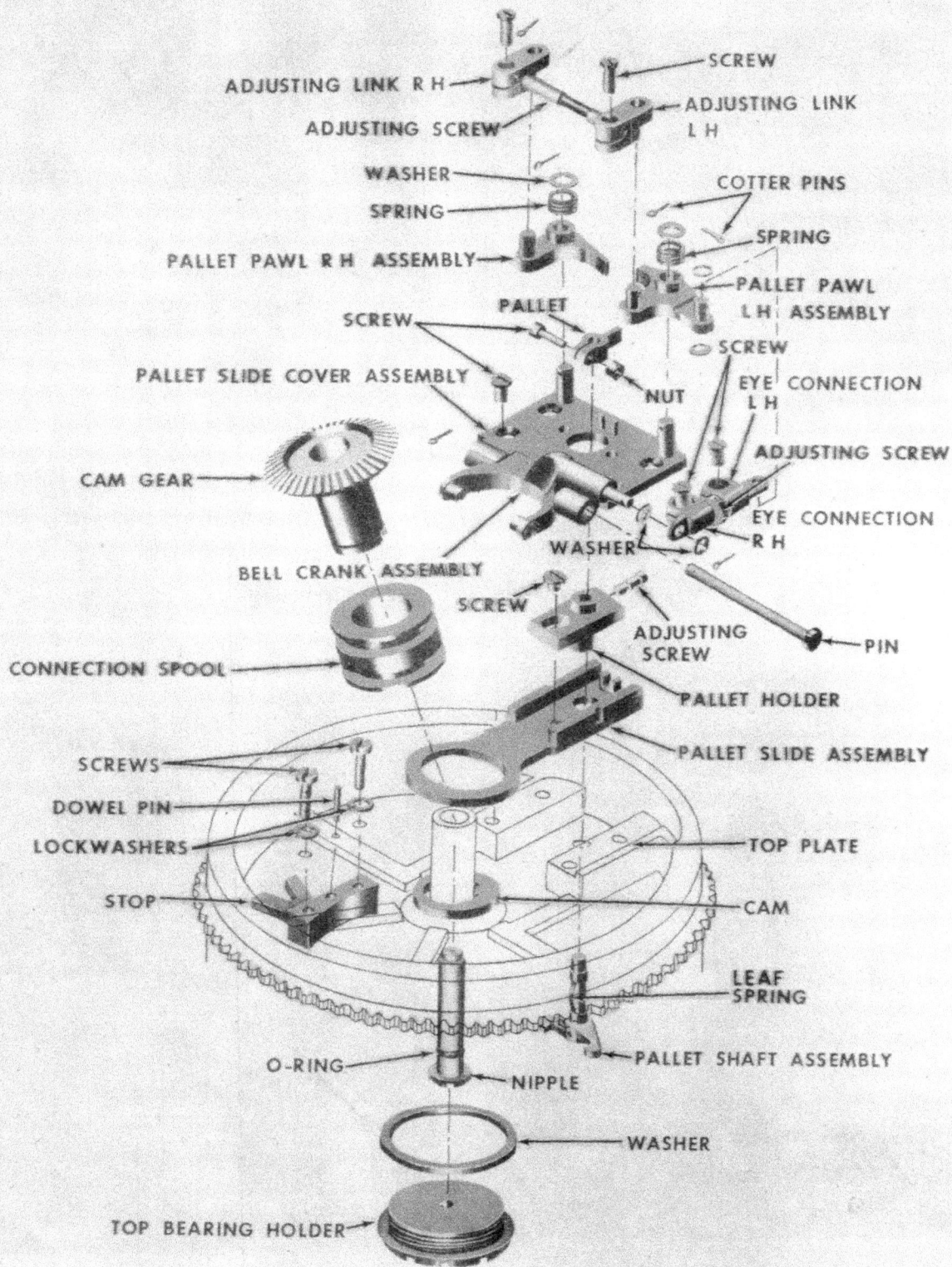

Figure 2-57. Pallet Mechanism.

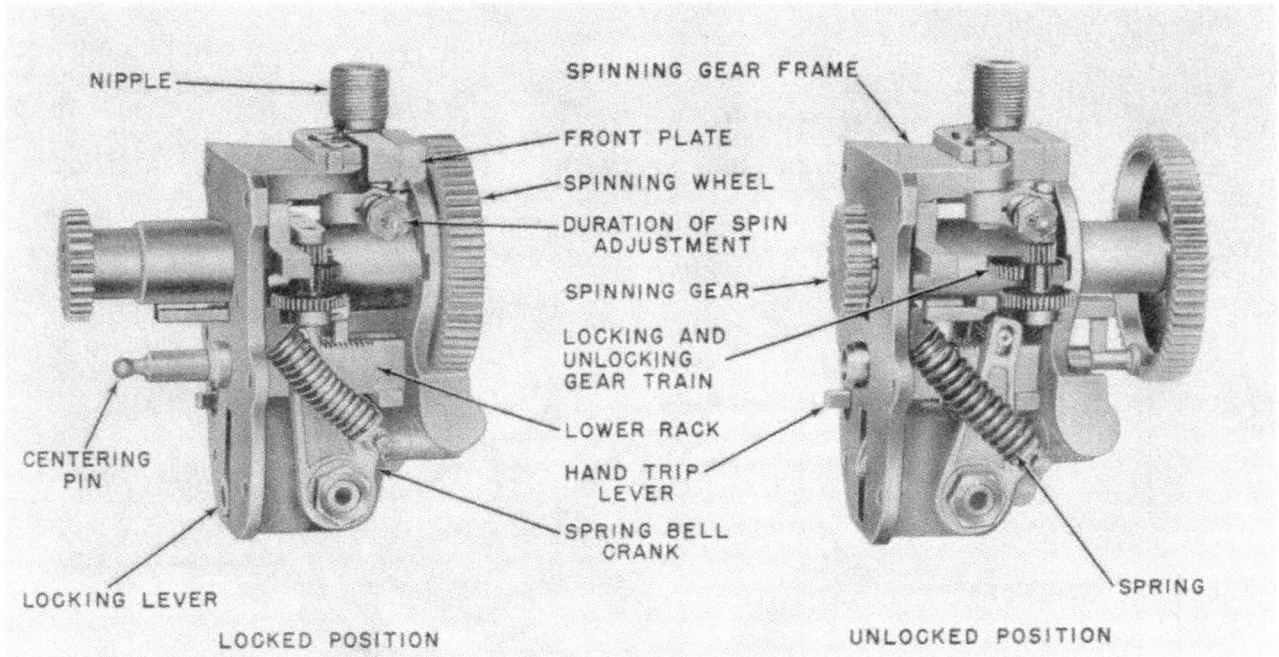

Figure 2-58. Spinning and Unlocking Mechanism.

Figure 2-59. Gyro in Locked Position.

pin, which prevents axial gyro movement until the gyro wheel is brought up to normal spinning speed. The centering pin is contained in the lower rack secured to the frame and front plate and has a ball on its end that inserts into the hole at the locking end bearing of the gyro inner gimbal. When in its locking position, the centering pin holds the spinning gear of the spinning wheel and impulse mechanism in mesh with the gear on the gyro wheel. Unlocking the mechanism withdraws the centering pin and spinning gear to unlock the gyro. Unlocking time is controlled by the duration-of-spin mechanism.

2-147. The spinning wheel, gear, and impulse mechanism is powered by full air-flask pressure to bring the spinning of the gyro wheel up to speed quickly. After 24 ±4 revolutions of the spinning wheel, the spinning gear and centering pin are withdrawn and air-flask pressure to the spinning wheel is shut off. Air-flask pressure is delivered to the spinning wheel through the impulse valve and nozzle ports in the lower portion of the front plate. The impulse valve is held open during the start of the torpedo run by a swivel block on the valve bell crank. The impulse of air-flask pressure rotates the spinning wheel attached to the spinning gear by a thrust-type sleeve and shaft. When the spinning operation is completed, the spinning and unlocking mechanism becomes unlocked (figure 2-58) and causes the spinning gear to unmesh from the gyro wheel gear, and the spinning wheel moves away from the nozzle ports in the front plate. At the same time, the spring-loaded bell crank moves in an aft position, permitting the impulse valve to seat, where it remains closed by air pressure.

2-148. **Bracket and Gearing Assembly.** The bracket and gearing assembly is mounted on the top forward side of the gyro pot (figure 2-53) and is secured to the immersion casing (figure 2-54). It transfers the drive from the forward propeller shaft pinion gear to the cam gear of the pallet mechanism on the gyro top plate assembly. It consists of the bracket and bearing assemblies to support the pallet driving gear and spindle, which are interconnected by a universal driving connection. The pallet driving gear is composed of a shaft with a smaller diameter central portion, a spur gear on one end, and a square socket in the other end. Its central portion revolves in a bearing assembly made up of two parts secured together with four screws. The aft end of the bearing assembly pivots on two trunnion pins secured in the bracket; the forward end rides on a spring-loaded piston secured in the bracket. The pallet driving spindle is composed of a shaft with a smaller diameter central portion, a bevel gear on one end, and a square socket in the other end. Its central portion revolves in a bearing with one portion an integral part of the bracket and the other portion a bearing cap secured to the bracket with six screws. The universal driving connection is a shaft with a combination round-and-square drive at each end that mates with the square sockets of

the pallet driving gear and spindle. The pallet driving gear is mounted in a floating bearing assembly to compensate for the play that would otherwise exist between the pallet driving gear and the forward propeller shaft pinion gear.

2-149. **Valve Rock Shaft Assembly.** The valve rock shaft assembly (figure 2-60), interconnects the gyro pallet mechanism and adjusting head shaft to position the pilot valve of the vertical steering engine. The adjusting head shaft is connected by a clamp to the valve rock shaft and by a flat link to the pilot valve of the vertical steering engine. The other end of the shaft rides in the upper groove of the pallet mechanism connection spool. The operation of the combined linkage transfers off-course signals from the pallet mechanism to the vertical steering engine. The valve rock shaft assembly has its shaft mounted on the bracket and gearing assembly and is held in place by a bearing cap and two screws. Its shaft pivots in its bearing and is provided with a counterweight to produce a steady transfer of off-course signals.

2-150. **Gyro Reducing Valve.** The gyro reducing valve is mounted on the starboard side of the immersion gear casing (figure 2-52) and decreases working air pressure as required to maintain a constant correct spinning speed of the gyro wheel. The working air pressure is delivered from a tee connector on the vertical steering engine and through a pipe to the elbow connection of the gyro reducing valve. When working air pressure enters the valve, it is forced against the spring-loaded valve and diaphragm to permit a restricted airflow into the space between the diaphragm and the valve body. The amount of airflow from the gyro reducing valve depends on the adjustment of spring tension on the valve stem. The adjusted amount of air pressure flows from one side of the tee connector on the valve body through a pipe to the gyro top plate nipple (figures 2-52 and 2-57) and through the top bearing holder and air passages in the gyro outer gimbal ring to buckets on the gyro wheel.

2-151. **Vertical Steering (Course) Engine.** The vertical steering engine is mounted on the starboard side of the gyro pot (figure 2-52) and is connected to the vertical rudders by the rudder rod, the rudder rod connection, and the rudder yoke. It receives steering orders from the valve rock shaft assembly, which is controlled by the pallet mechanism, as influenced by the gyro off-course signals; it produces the force necessary to position the vertical rudders accordingly. It is a full-throw type of engine with a design that permits only hard-over rudder positions in accordance with the changes in off-course signals. It operates on the working air pressure supply received from the afterbody test connection. The vertical steering engine consists of a body with two cylindrical chambers, a piston, a pilot valve, three pipe connections, and two yoke connections. The larger of the two body chambers houses the piston. The smaller chamber houses the pilot

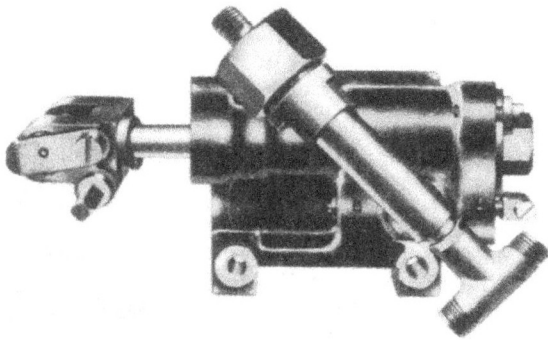

Figure 2-60. Vertical Steering Engine.

valve. Two small port holes interconnect the two chambers. Two exhaust ports are provided for the smaller (pilot valve) chamber. The piston has a larger diameter with an O-ring that rides on the walls of the larger chamber, an adjusting washer to regulate its amount of throw, and a shaft that rides in a body hole with an O-ring to prevent air pressure from escaping from that portion of the body. The pilot valve rides in a sleeve and has grooves to direct input air pressure through sleeve holes and body port holes to one or the other end of the piston and to direct exhausted air pressure through the body exhaust ports. The input air pipe connection is equipped with an air strainer and holder. The output air pipe tee connector directs air pressure to the distance valve of the enabler and the gyro reducing valve. The yoke connection on the pilot valve is connected by a linkage to the valve rock shaft assembly. The yoke connection on the piston shaft is connected to the vertical rudder rod.

2-152. Deleted by CHANGE 2

2-153. Gyro Setter Assembly. The gyro setter assembly is a portion of the gyro angle order servo system. It is mounted on top of the gyro pot of the control mechanism (figure 2-51). It is wired to the R-C network in the afterbody through the afterbody cabling and torpedo A- and B-cables to the submarine fire control system. It positions the gyro top plate in accordance with the gyro angle orders (G) transmitted from the computer of the fire control system. Its major components are two synchro transmitters SG1 and SG2, servomotor M1, gyro setter brake solenoid S1, and a gear train, all of which are mounted on the same bracket assembly.

2-154. Immersion Mechanism. The immersion mechanism consists mainly of the horizontal steering engine (figure 2-52), the air chamber and diaphragm assembly (figure 2-54), the pendulum and casing assembly (figure 2-55), and the depth-setter assembly. Most of the immersion mechanism components are installed on the forward area of the gyro pot and base assembly.

2-155. Pendulum and Casing Assembly. The pendulum and casing assembly (figure 2-55) is one of the major components of the immersion mechanism. The immersion gear casing and gasket are bolted to a flange over a circular chamber in the forward end of the gyro pot and base assembly and forms a framework for the mounting of the pendulum and depth spring assembly. The casing is Y-shaped with a hollow cylindrical bottom and two upper arms. Part of a knife-edge bearing is secured in a hole at the top of each arm and protrudes outward to support the other part of the knife-edge bearings in the top of the pendulum arms. An external nipple of the hollow cylindrical bottom is connected by a pipe assembly to a connection on the afterbody shell to allow entry of sea-water pressure to the diaphragm. Each of two rollers are secured by a screw to the casing base and at each side of the hollow bottom to aline the swinging motion of the pendulum. The pendulum is made up of two sections for convenience of assembly. Its main (forward) section has a base filled with lead and two suspension arms by which the pendulum hangs. Its lead-filled after section is attached to the main section by screws and nuts. The lead is added to give the pendulum adequate mass. When the two sections are assembled, they encircle the bottom portion of the immersion gear casing. Fore and aft swing of the pendulum is limited by stops on the immersion gear casing. Retainer plates are secured to the top outside of the pendulum arms to prevent the knife-edge bearings from being damaged by jumping and pounding forces applied in a vertical direction during torpedo run and handling. When torpedo roll and handling exceeds 30° to either side, guides installed on the pendulum underside bear on the rollers installed on the bottom of the immersion gear casing to prevent damage that would otherwise be inflicted by forces applied in a horizontal direction.

2-156. Buffer Springs and Tension Rod Assembly.
The buffer springs and tension rod assembly (figure 2-52) is mounted on the lower inner side of the left-hand pendulum arm. In conjunction with the valve operating lever, it transfers pendulum movement to the pilot valve of the horizontal steering engine. It also cushions any quick change in pendulum movement before it is transferred to the pilot valve. It consists of a rod, a pivot bearing, two springs, four buttons, a nut, a connection eye, and two cotter pins. The rod has a larger central diameter that rides in the pivot bearing attached to the pendulum arm. Each of the two coil springs are mounted between two buttons, all installed on each of the two smaller rod diameters. One spring and two buttons are secured on the forward end of the rod with a nut and cotter pin. The other spring and two buttons are secured on the afterend of the rod with the connection eye and cotter pin. The connection eye is connected to the lower end of the valve operating lever.

2-157. Depth Spring Assembly. The depth spring assembly (figure 2-55) is housed in the hollow cylindrical bottom of the immersion gear casing. It reacts to sea-water depth pressure applied to the diaphragm of the air chamber. The movement of the depth spring is transferred from the depth spring and diaphragm lever to the pendulum lever. One end of each lever is locked to the same shaft. The other end of the depth spring and diaphragm lever acts as a fulcrum for both levers. The center of the depth spring and diaphragm lever is connected to the depth spring and diaphragm to receive the movement caused by changes in sea-water depth pressure. The other end of the pendulum lever is connected to the left-hand pendulum arm by an adjustable linkage and to the immersion gear casing by an auxiliary depth spring. An adjusting screw is inserted through a hole in the top of the hollow cylindrical portion of the immersion casing and screws into the top of the depth spring. The adjusting screw has a gear on its top end to mesh with the gear train of the depth setter which controls the position of the depth spring in relation to sea-water pressure at a preset depth order.

2-158. Air Chamber and Diaphragm Assembly.
The air chamber and diaphragm assembly (figure 2-54) is installed in the bottom forward end of the gyro pot and base assembly and seals the hollow cylindrical bottom of the immersion gear casing. Its major components consist of the air chamber, the diaphragm, the diaphragm ring, and a gasket. When the air chamber and the diaphragm are installed, the gasket is placed between the immersion gear casing and the flange of the gyro pot and base assembly; the diaphragm ring is secured to the inner bottom face of the same flange; the diaphragm and the upper face of the air chamber flange are secured on studs on the bottom face of the diaphragm ring; and the center of the diaphragm is secured by a plate and nut to the bottom of the depth spring. The diaphragm is made of an oil-resistant rubberized fabric.

2-159. Auxiliary Depth Spring. The auxiliary depth spring is a coil spring connected between the upper end of the left-hand arm of the immersion gear casing and the upper end of the pendulum lever (figure 2-52). It increases the reliability of a shallow running depth by applying a steady tension to the pendulum lever and associated linkages.

2-160. Climb-Angle Limiter Device. The climb-angle limiter device consists of an angular steel plate and an adjustable coil spring that form a breakaway type link holding the upper and lower portions of the pendulum lever together (figure 2-61). The angular steel plate is secured to the upper pendulum portion by two rivets and to the lower pendulum portion by a pivot screw and nut. The adjustable coil spring is mounted on an eyebolt attached to the lower pendulum portion, and its tension bears on the after edge of the angular steel plate.

2-161. Transportation and Replacement Screws.
The transportation and replacement screws are installed adjacent to the air chamber in a threaded hole of the diaphragm ring and insert into a hole in the base of the immersion gear (figure 2-54). The transportation screw is the longer of the two screws so that it engages a socket in the pendulum base to prevent pendulum movement during torpedo handling and shipping. Prior to torpedo launching and testing, the transportation screw must be removed and the replacement screw must be installed together with its O-ring to permit the pendulum to swing freely and to insure watertightness. A test pin made to closer tolerance than the transportation screw, is used to center pendulum when testing depth mechanism.

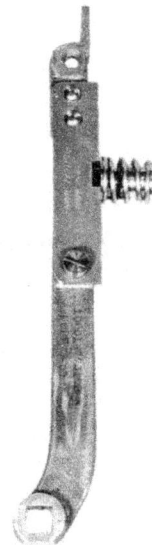

Figure 2-61. Climb-Angle Limiter Device.

2-162. Horizontal Steering (Depth) Engine. The horizontal steering engine (figure 2-62) is mounted on the port side of the gyro pot (figure 2-52) and is connected to the horizontal rudders by the rudder rod, the rudder rod connection, and the rudder yoke. It consists of a cylindrical body housing a piston that contains a pilot valve. The piston has two shafts, one riding in a bearing hole in the cylinder body aft end and the other riding in the bearing hole of a plug installed in the cylinder body forward end. The central portion of the piston contains one-way vent ports for distribution of working air pressure to each end of the piston and two O-rings to make an airtight bearing surface between the piston and the cylinder body. Both piston shafts and central portions are hollow to accept the pilot valve and a threaded stop-plug. The pilot valve has bearing land surfaces and vent holes between the land surfaces. At the forward end of the pilot valve is an adjustable screw type link that connects to the valve operating lever connected to the buffer springs and rod assembly. At the end of the after piston shaft is a yoke that connects to the rudder rod. An air strainer is installed on an external threaded nipple (input) of the cylinder body.

2-163. Depth Setter Assembly. The depth setter assembly is a portion of the depth-set order servosystem. It is mounted on the forward side of the immersion gear casing (figure 2-51). It is wired to the R-C network in the afterbody and through the afterbody cabling and torpedo A- and B-cables to the fire control system. It sets the correct tension on the depth spring in accordance with the depth-set

order (HQZ) transmitted from the computer of the fire control system. Its major components are synchro transmitter SG3, servomotor M2, and a gear train, all of which are installed on a cast aluminum bracket.

2-164. IDLER GEAR AND BRACKET ASSEMBLY. The idler gear and bracket assembly (figure 2-63) is doweled and bolted to the after side of the afterbody vertical bulkhead. It is driven by the pinion gear on the forward propeller shaft and transmits its drive directly to the governor assembly through a solid shaft with universal ends. It also indirectly drives the enabler assembly via a flexible shaft connected to the governor assembly. It consists of three gears, two shafts, and a bracket. Two of the gears are secured to one shaft, which is free to rotate in a bracket bearing hole. The remaining gear is secured to the other shaft, which is free to rotate in another bracket bearing hole. The outside one of the two gears meshes with the pinion gear on the forward propeller shaft, while the inside gear meshes with the remaining gear, which has a square hole in its outer center to accept the universal end of the solid shaft to drive the governor assembly.

2-165. GOVERNOR ASSEMBLY. The governor assembly (figure 2-64) is installed in the oval flange at the top center of the afterbody (figure 2-33) and shuts down the torpedo if the main engine overspeeds. It consists mainly of the governor, a tripping lever, and an overspeed switch mounted on a frame. The governor is mounted in bearing surfaces formed at the ends of two yoke extensions at

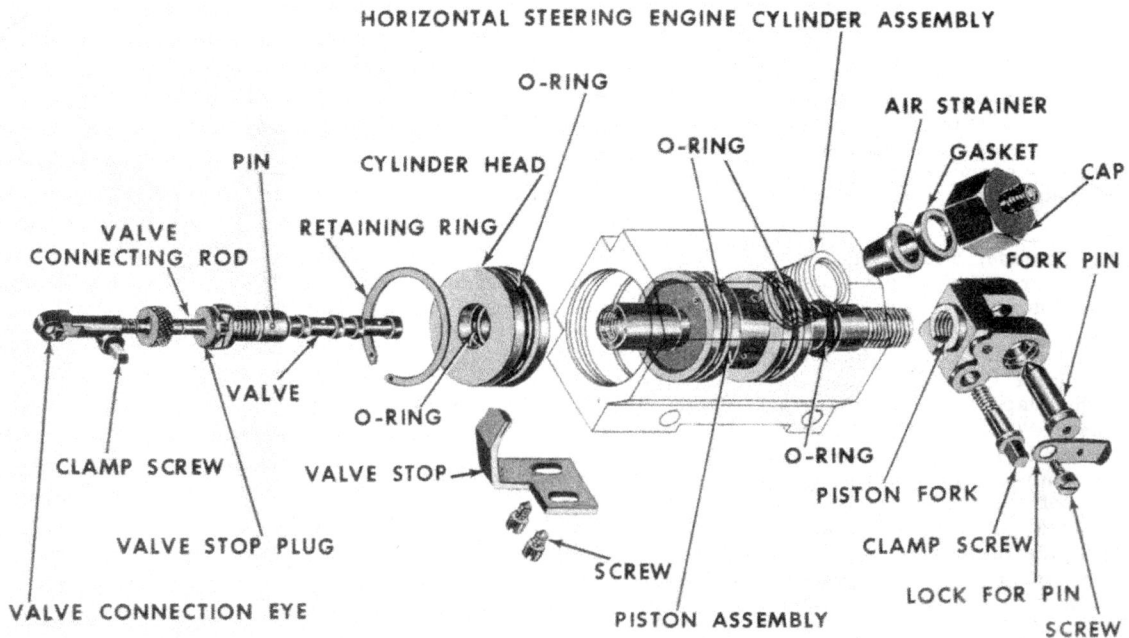

Figure 2-62. Horizontal Steering Engine, Exploded View.

the bottom of the frame and by two journal caps attached to the yoke extensions to allow the governor to rotate freely. The tripping lever and overspeed switch are attached to a common mounting bracket secured to one of the yoke extensions and adjacent to the governor. The governor has a rectangular body with curved sides and a shaft protruding from each end and contains two spring-loaded weights. The two weights are seated in bored openings at opposite sides of the governor body and are held in place by two coil springs. The tripping lever has two pawls, each secured to opposite ends of the shaft that rides in a bearing hole in the mounting bracket. The overspeed switch is a sealed microswitch with an operating lever and has a two-wire cable wired across its normally closed contacts. An adjusting screw holds the relative pawl and switch operating lever together in a position to keep the overspeed switch open. When the torpedo main engine is running at normal speed, the pinion gear on the forward propeller shaft drives the idler gear and bracket assembly, and eventually the governor body, at a low enough rate to allow the spring-loaded weights to remain fully seated in the governor body. However, when the torpedo main engine develops excessive speeds, driving the governor at a rate of 2200 to 2250 rpm, centrifugal force of the governor-body spinning partially overcomes the spring tension on the weights, causing them to protrude. The amount of protrusion is the same for both weights, because the spring tension is the same on each weight. Both weights protrude equally, to maintain the balance required for smooth governor operation with minimum vibration. When the speed is great enough, the forward spring-loaded weight trips the lower pawl of the unlocking lever. Consequently, the upper pawl trips the operating lever of the overspeed switch to close the applicable portion

of the power pack circuit required to actuate the Navol shutoff valve.

2-166. Deleted by CHANGE 2

2-167. AFTERBODY ELECTRICAL CABLING. The afterbody electrical cabling (figure 2-66) is divided into the following three units:

1. Afterbody Receptacle and Harness Assembly.
2. Control Mechanism Wiring Harness.
3. Navol Surveillance and Power Cable.

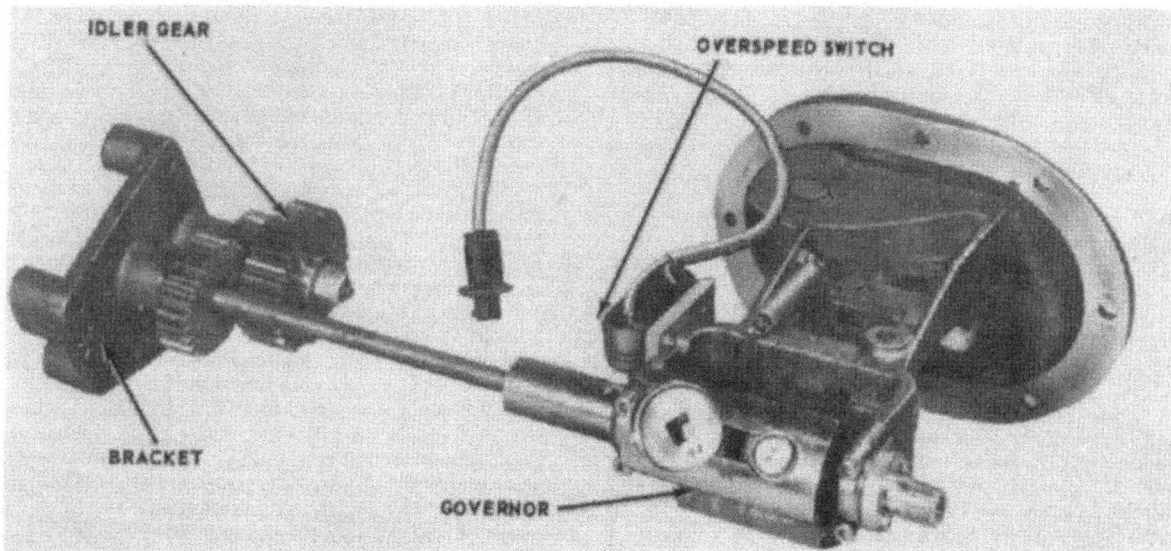

Figure 2-63. Idler Gear, Bracket, and Governor Assembly.

Figure 2-64.   Governor Assembly.

Figure 2-65.   Deleted by CHANGE 2

2-168. Afterbody Receptacle and Harness Assembly. The afterbody receptacle and harness assembly receives signals from the submarine fire control system through torpedo A- and B-cables and distributes the signals to the proper servosystem components in the afterbody. Navol decomposing rate signals from the monitoring unit are transmitted through leads of the Navol surveillance and power cable to the afterbody receptacle PG, then carried by various connectors to the indicator panel. In a reverse direction, the receptacle and harness assembly also routes the 115-volt 60-Hz firing power from the fire control system through leads of the Navol surveillance and power cable to the power pack in the energy section to start the torpedo.

2-169. The afterbody receptacle and harness assembly consists of 65-conductor receptacle PG, the main harness, and three secondary harnesses branching from the main harness. One end of the main harness terminates in receptacle PG, which receives the mating connector of the torpedo control A-cable. The other end of the main harness terminates in connector SE1, which receives mating connector PE1 of the control mechanism wiring harness. One of the secondary harnesses terminates in connector SE9, which mates with connector PE9 of the R-C network. Another secondary harness terminates in connector PE6, which mates with connector W1P3 of the Navol surveillance and power cable.

2-170. Receptacle PG is installed in a flange in the bottom and near afterend of the afterbody shell (figure 2-33). An O-ring is installed on the outer surface of the receptacle to provide a watertight joint on installation. Three screws secure the receptacle in the flange. A blanking cover with an O-ring is installed in the receptacle to provide protection when the receptacle is not in use. The receptacle has a bayonet locking arrangement for securing the protective cover and torpedo control A-cable connector in place. All the wires of the afterbody receptacle and harness assembly are Teflon-insulated for oil and heat resistance and are bundled in appropriate branches by lacing with silicone-coated cord.

2-171. Control Mechanism Wiring Harness. The control mechanism wiring harness consists of various leads and connectors to join the gyro setter and the depth setter electrical components through the afterbody receptacle and harness assembly and torpedo control A- and B-cables to the fire control system. The main branch of the harness starts at connector PE1, which mates with connector SE1 of the afterbody receptacle and harness assembly. A set of auxiliary branches has wire leads that connect to synchro transmitters SG1 and SG2, servomotor M1, and solenoid brake S1 of the gyro setter. Another set of auxiliary branches has wire leads that connect to synchro transmitter SG3 and mating connectors SE10 and PE10, through which wire leads are connected to servomotor M3. At each synchro transmitter and servomotor, the harness leads are provided with silicone rubber caps. When the leads are attached to the synchro transmitters and servomotors, all caps are packed with grease and are coated externally with a neoprene coating. All wires of the control mechanism wiring harness are Teflon-insulated for oil and heat resistance and bundled in appropriate branches by lacing with silicone-coated cord.

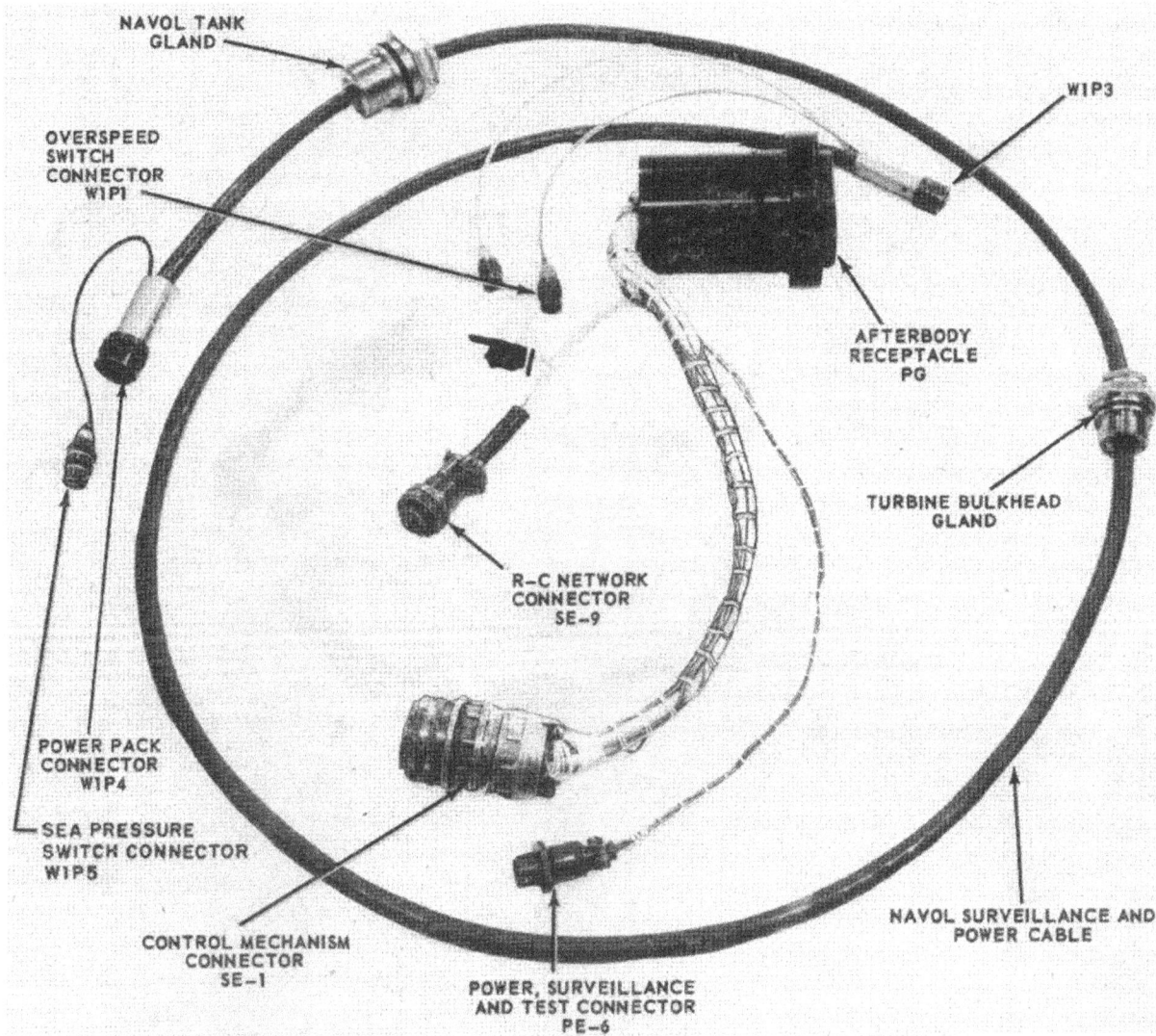

Figure 2-66. Afterbody Electrical Cabling and Connectors.

2-172. Navol Surveillance and Power Cable. The
Navol surveillance and power cable interconnects
the monitor unit and power pack in the Navol tank
and valve compartment of the energy section with
the afterbody receptacle and harness assembly and
the overspeed switch of the governor assembly in
the afterbody. Its main portion is encased in silicone shielding and has connector W1P3 at its free
end and extends from the dry section of the Navol
tank and valve compartment. Two leads, encased
in Teflon sleeving, extend from back of connector
W1P3, terminating at connector W1P1. Two other
leads, also encased in Teflon sleeving, extend from
the back of connector W1P4 and terminate at connector W1P5. The exterior of the main portion of
the Navol surveillance and power cable has two

gland fittings to seal the cable installation in the
gland openings of the afterbody turbine bulkhead
and the bulkhead between the wet and dry sections
of the Navol tank and valve compartment. Connector W1P1 mates with connector 1A1P1 of overspeed
switch S1 on the governor assembly. Connector
W1P3 mates with connector PE6 of the afterbody
receptacle and harness assembly. Connector W1P4
mates with connector 2A3P6 on the power pack.
Connector W1P5 mates with the plug of sea-water
pressure switch S4 in the Navol tank and valve compartment.

2-173. AFTERBODY PIPING. There are several
pipe assemblies in the afterbody. Most carry reduced air pressure (working air pressure 635 psi

nominal); two carry high-pressure air (air flask pressure 2800 ±50 psi), and one carries sea-water depth pressure. There are two nonpressure pipes permanently attached to the afterbody after bulkhead; these carry sea water from the scoops on the aft end of the afterbody shell to the cooling system for the after bulkhead bearing. The pipe assemblies are listed in the following paragraphs.

2-174. Low-Pressure Air. This piping includes the:

1. Turbine bulkhead to engine casing bracket.

2. Engine casing bracket to test connection.

3. Test connection to horizontal steering engine.

4. Test connection to vertical steering engine.

5. Vertical steering engine to distance valve.

6. Deleted by CHANGE 2

2-175. High-Pressure Air. This piping includes the:

1. Turbine bulkhead to engine casing bracket.

2. Engine casing bracket to spinning and unlocking mechanism.

2-176. Sea Water Depth Pressure. This piping runs from the water inlet connection to the immersion gear casing.

2-177. AFTERBODY EXPENDABLES AND FIR PARTS. All torpedo control A-cable components are discarded after use in a warshot or exercise shot. The remaining expendables are removed, treated, and shipped to a designated refurbishing activity after use in an exercise shot. The used FIR parts, 2a through 2e, are packed in the Navol tank and valve compartment for return to the FIR activity. FIR parts 2f through 2l should be returned to a FIR activity if they do not function properly or fail to pass prescribed tests. However, they shall be returned to the FIR activity every three years. When the torpedo is fired, the afterbody contains the following expendables and FIR parts:

1. Expendables.

   a. Torpedo Control A-Cable Mk 1 Mod 12 or Mod 13.

   b. Two Igniters Mk 6 Mod 4.

   c. Catalyst Cartridge.

2. FIR Parts.

   a. Navol Restriction.

   b. Fuel and Water Sprays.

   c. Four Exhaust Valves.

   d. Decomposition Chamber O-ring.

   e. Spray Gaskets.

   f. Gyro.

   g. Deleted by CHANGE 2

   h. Gyro Setter.

   i. Depth Setter.

   j. Spinning and Locking Mechanism.

   k. Depth Engine.

   l. Steering Engine.

2-178. Torpedo Control A-Cable. Either one of two torpedo control A-cables (Mk 1 Mod 12 or Mk 1 Mod 13) (figure 2-67) can be used in firing the torpedo. In their application for this torpedo, the difference in cable length and connector wiring between the two cables produce no undesirable effect. It is recommended that all torpedo control A-cables Mk 1 Mod 12 be used until their supply is exhausted before using a torpedo control A-cable Mk 1 Mod 13. (Refer to OP 2639 for additional information on the application of the two torpedo control A-cables.) Each torpedo control A-cable consists of a 65-conductor shielded cable (69 or 89 inches in length) with watertight connectors on each end. One end is installed in the receptacle mounted on the inside of the tube door and the other in the torpedo afterbody receptacle. The connector used for tube-door connection is equipped with two handles (one contains a fuse), a spring-type latch to lock the connector in the tube door receptacle, and an external O-ring to make the installation watertight. The connector used for torpedo connection is equipped with a cable cutter, a bayonet-type socket arrangement to lock the connector in the afterbody receptacle, and an O-ring to make the installation watertight. The cable cutter (figure 2-68) is attached to the connector installed in the afterbody receptacle and is held in place by a pivot screw and shear pin. The cable is secured in a channel of the cutter by a locking wire. When the torpedo is launched, cable slack is tightened and the resulting tension is transferred to the cutter. When the tension becomes great enough, a rotary motion is applied to the cutter in an aft direction to shear the shear pin and cable close to the exterior surface of the connector; then the sheared end of the cable is pulled free from the cutter channel; the lock wire may or may not be broken.

2-179. Igniters. The two igniters Mk 6 Mod 4 are described in paragraph 2-115. They are shipped in sealed reusable containers and shall remain sealed in their individual containers until ready for installation into the torpedo. All igniters that are expended or fail to operate shall be shipped in a reusable container to Naval Torpedo Station, Keyport, Washington, in accordance with NAVORD Instruction 8510.10.

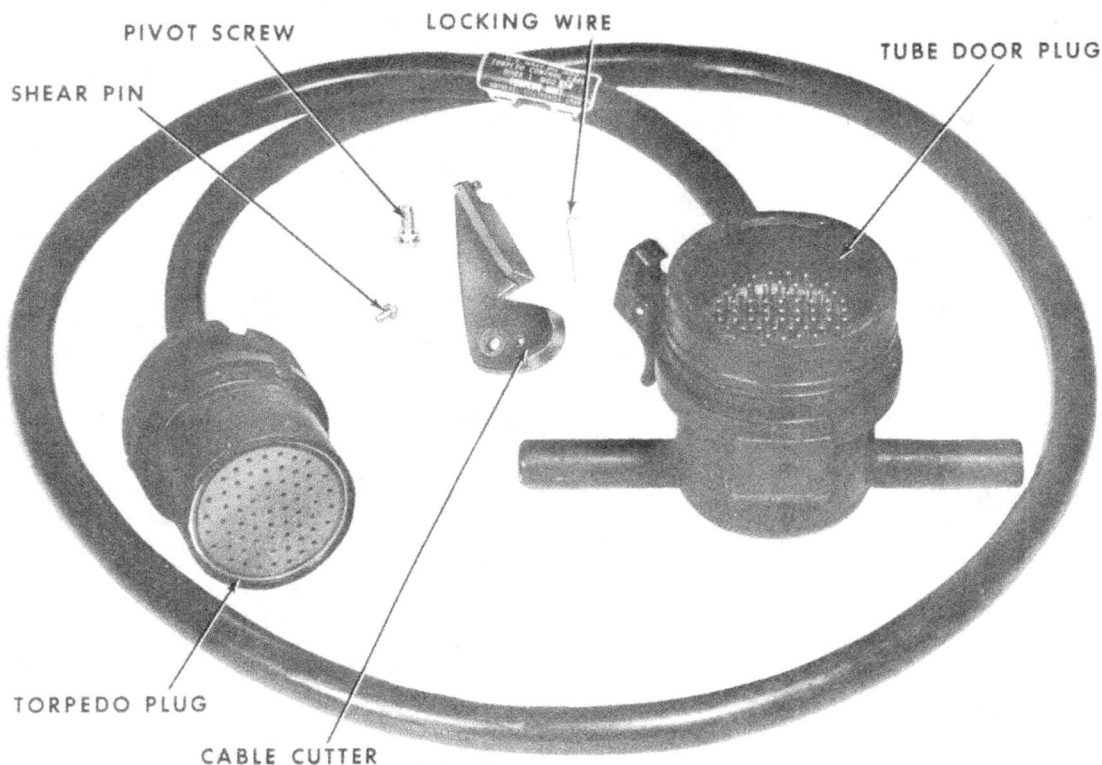

Figure 2-67. Torpedo Control A-Cable Mk 1 Mod 12.

2-180. Catalyst Cartridge. The catalyst cartridge is described in paragraph 2-105. It is shipped in a sealed reusable container. An expended catalyst cartridge shall be disassembled and its refurbishable components shipped in a reusable container to the nearest supply activity in accordance with SPCC Instruction 4440.83.

2-181. Navol Restriction Assembly. The Navol restriction assembly is described in paragraph 2-107. It is a FIR part shipped in a specially wrapped package included with the Navol tank and valve compartment shipment. A used Navol restriction assembly shall be included in the return shipment of the Navol tank and valve compartment in accordance with instructions prescribed in Volume 2.

2-182. Fuel and Water Sprays. The fuel and water sprays are described in paragraphs 2-108 and 2-112. They are FIR parts shipped in specially wrapped packages included with the Navol tank and valve compartment shipment. Used fuel and water sprays shall be included in the return shipment of the Navol tank and valve compartment in accordance with instructions prescribed in Volume 2.

2-183. Exhaust Valves. The four exhaust valves are described in paragraph 2-94. They are FIR parts shipped in specially wrapped packages included with the Navol tank and valve compartment shipment. Used exhaust valves shall be included in the return shipment of the Navol tank and valve compartment in accordance with instructions prescribed in Volume 2.

2-184. TAIL ASSEMBLY. The tail assembly (figures 2-69 and 2-70) is assembled to the aft end of the afterbody and completes the afterbody and tail section of the torpedo. It consists of torpedo stabilizing, steering and propelling modules made up of the following major components:

1. Tail Cone.

2. Tail Blades.

3. Rudders and Linkages.

4. Propellers and Mountings.

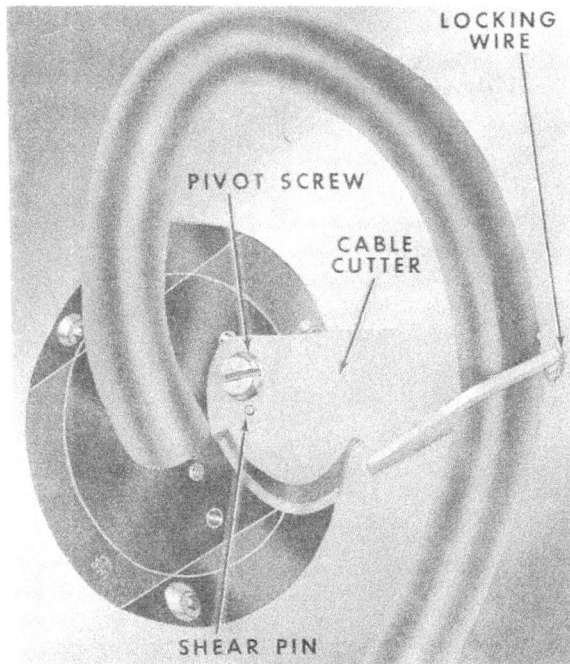

Figure 2-68. Cable Cutter.

2-185. Tail Cone. The tail cone is the shell of the tail assembly and provides for installing the blades, rudders, and linkages (figure 2-71) and the propellers and mountings (figure 2-72). There are four longitudinal slotted fins 90° apart on the external surface of the tail cone to which the blades are riveted as permanent parts. The forward end of the tail cone contains a wedge-type joint around its internal perimeter and 16 counterbored holes equally spaced around its outer perimeter for securing the tail cone to the afterbody after bulkhead. Unlike the other wedge-type joints of the torpedo sections, this joint is more of a butted type and does not contain an O-ring for watertightness. With the mated joints butted together, the tail cone is secured to the after bulkhead by 16 securing screws installed in the oblique holes of the tail cone and tapped holes in the after bulkhead. Near the aft end of the tail cone, adjacent to each rib and blade, is a hole to accommodate the inner bearing for installing a horizontal or vertical rudder. The forwardmost pair of opposite inner bearing holes support the horizontal rudders and their linkage. The aftermost pair of opposite inner bearing holes support the vertical rudders and their linkage. Around the face of the tail cone afterend are 12 tapped holes to receive securing screws that hold the tail bearing, which supports the afterend of the propeller mountings. A series of graduations are engraved on the exterior surface of the tail cone, aft of the port horizontal rudder installation. The graduations are used in calibrating linkage adjustments to position the horizontal rudders in an angular plane that agrees with orders from the horizontal steering (depth) engine. A zero line is provided on the aft starboard exterior surface of the

tail cone to provide a reference for checking the top surface alinement of the horizontal rudders. No calibrating graduation is provided for checking the vertical rudder position, because the vertical rudders assume only hard-over left and right positions that agree with orders from the vertical steering (course) engine. Vertical rudders are checked by insuring that the rudders move equally left and right when the vertical steering engine is operated. There are four counterbored and threaded holes through the tail cone shell near the forward bottom portion of the tail cone and two at each bottom quadrature of the tail cone. They are used to gain access to the rudder linkages and to drain residues from the tail cone. When not in use, each is blanked off by a plug and washer.

2-186. Tail Blades. The tail blades are rigidly mounted sheet-steel vanes that help stabilize the torpedo during a run. They extend forward beyond the tail cone and overlap the aft end of the afterbody. The horizontal blades are located forward of the vertical blades at a distance equal to the difference between the wider horizontal rudders and narrower vertical rudders. For additional support, the forward end of each horizontal blade is secured with screws in a bracket on the afterbody shell. The slight forward extension of the vertical blades does not require additional support.

2-187. Rudders and Linkages. The horizontal rudders are connected by a linkage to the horizontal steering (depth) engine. The vertical rudders are connected by a linkage to the vertical steering (course) engine. The lower vertical rudder is cut away and has less area than the upper vertical rudder to minimize heel during turns. Both horizontal and vertical rudders are supported by bearing spindles near their leading edge. The outer bearing spindles are supported in outboard bearings attached to the trailing outer edge of each respective tail blade. The outboard bearings also serve as guides for installing the afterend of the torpedo in the launching tube. Because the diameter of the tail assembly is smaller than the tube, the outboard bearings prevent misalinement and binding when the torpedo passes through and out of the launching tube. The inner bearing spindles of each pair of horizontal and vertical rudders are interconnected by a C-shaped yoke. Each yoke is connected by an adjusting rod to rudder rod and steering engine connections in the afterbody. The afterend of each adjusting rod pivots on a stud attached to the related yoke. The forward end of each adjusting rod consists of a threaded clamping arrangement that contains an adjustable eyebolt. The eyebolt is secured in place by a screw that tightens the threaded clamping arrangement. Each half-turn of the eyebolt produces a longitudinal adjustment equal to one-half the thread pitch. When the rudders are actuated, the C-shape of the yokes and the offset position of the adjusting rods change the longitudinal movement into rotary motion and prevent interference with the propeller sleeves. When the torpedo run becomes stable, the torpedo travels at its set depth and course but with a tendency to assume a noseup attitude. A certain amount of down position of the

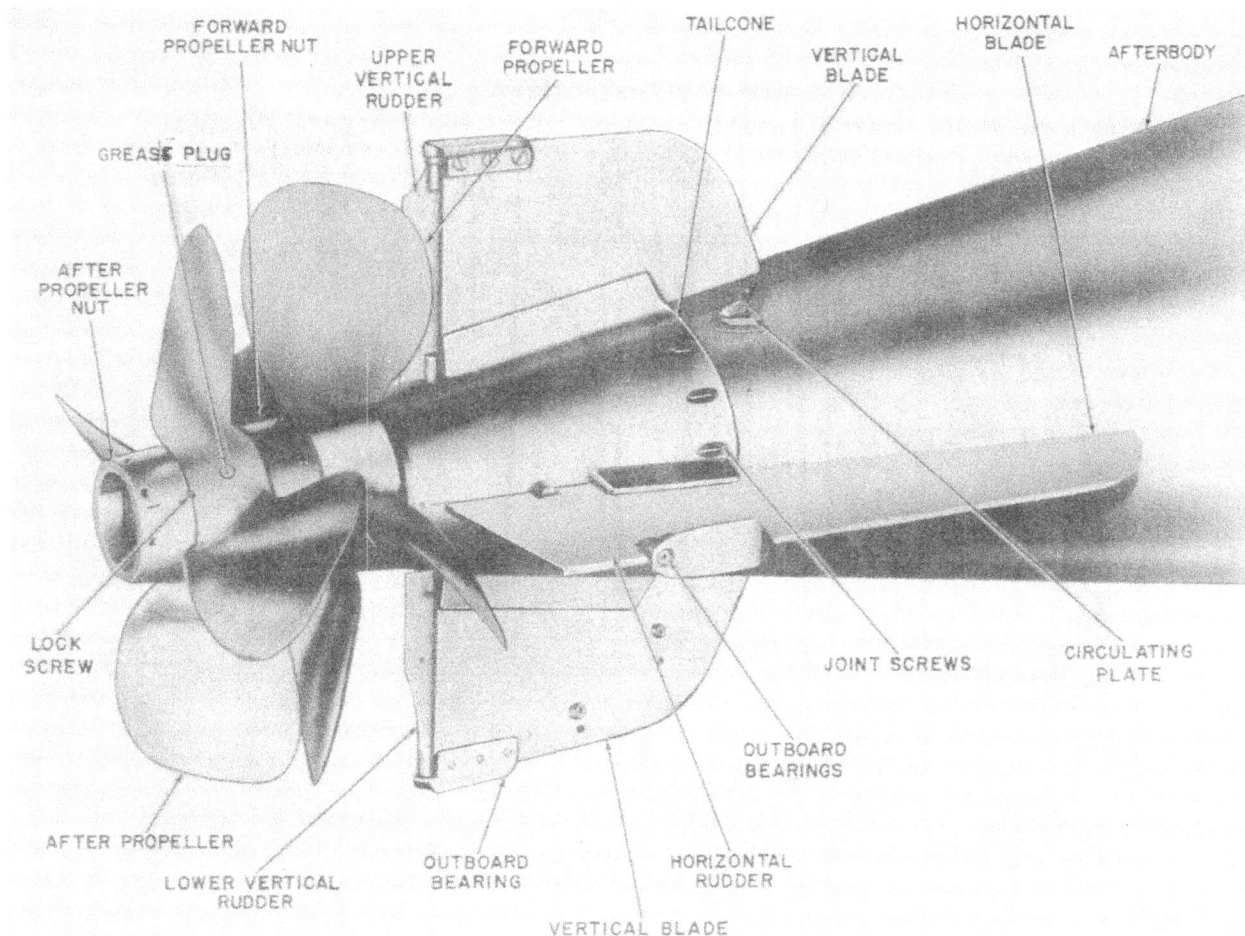

Figure 2-69. Tail Assembly, Starboard Rear View.

horizontal rudders is necessary to maintain proper torpedo attitude as required to keep the torpedo at set depth. The overall range of the horizontal rudder positions is 5 units (1 up and 4 down) of the graduated scale engraved on the tail cone. Each unit is an arbitrary measurement of horizontal rudder deflection. The measurement is taken from the 0 position on the scale, which is in alinement with the top surface of the port horizontal rudder. However, the 0 position is not the neutral position, which is 1.5 down.

2-188. Propellers and Mountings. The propellers and their mountings are installed in the afterend of the tail cone. There are two (one forward and one after) propellers. The forward propeller has four blades with a right-hand pitch and rotates clockwise (looking forward). The after propeller has four blades with a left-hand pitch and rotates counterclockwise. The opposite propeller rotations provide a balanced forward thrust with practically no tendency to create torpedo roll.

2-189. Both propellers have a tapered hole in their hubs. The hub of the forward propeller is mounted on a separate split hub with a straight hole that fits snugly on the forward propeller sleeve with a single key. The split hub has a single key on its tapered bearing surface for the hub of the forward propeller. The forward propeller and split hub are secured to the forward propeller sleeve by the forward propeller nut, which has a left hand thread to prevent loosening due to torque. The hub of the after propeller is mounted snugly on the tapered afterend of the after propeller sleeve and is secured in place by a key and the after propeller nut. Setscrews lock both propeller nuts in place.

2-190. Both propeller sleeves are hollow and contain several holes near their forward ends to allow exhaust gases to pass into the sea. The after propeller sleeve fits into the forward propeller sleeve and four grease bushings support the opposite rotation of the two sleeves. The forward (outer) propeller sleeve is keyed to the forward propeller shaft

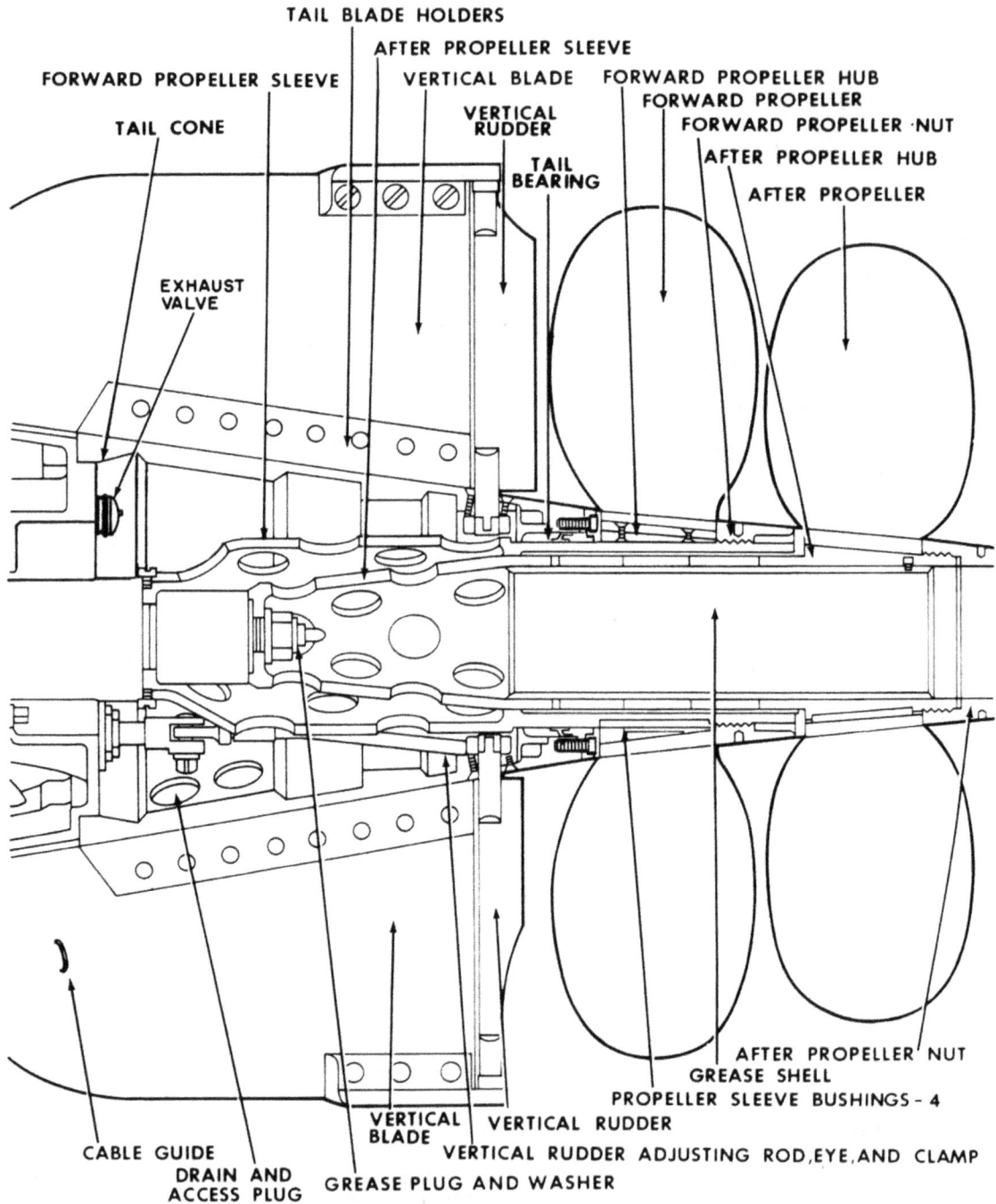

Figure 2-70. Tail Assembly, Sectional View.

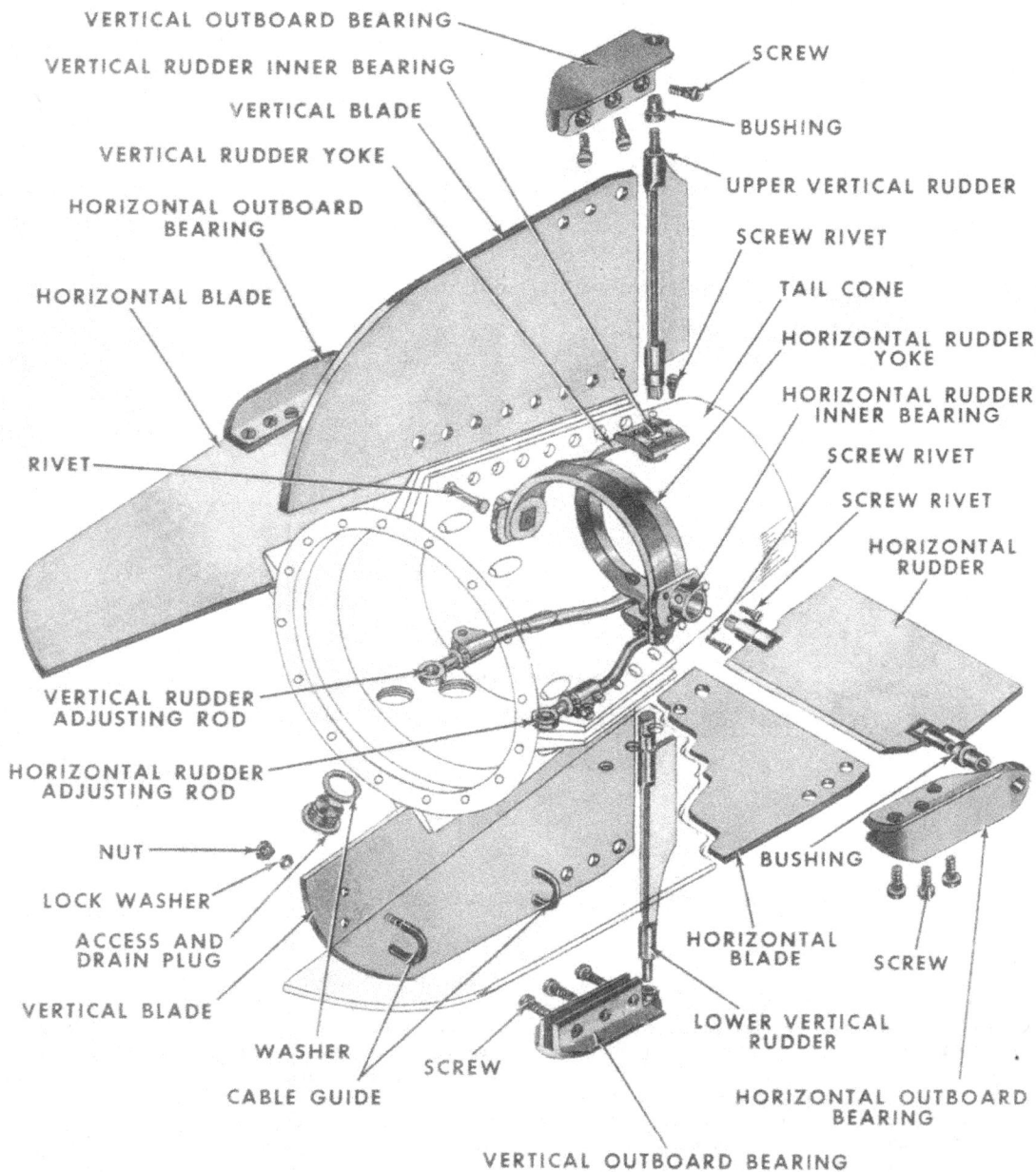

VERTICAL OUTBOARD BEARING

VERTICAL RUDDER INNER BEARING

VERTICAL BLADE

VERTICAL RUDDER YOKE

HORIZONTAL OUTBOARD BEARING

HORIZONTAL BLADE

RIVET

VERTICAL RUDDER ADJUSTING ROD

HORIZONTAL RUDDER ADJUSTING ROD

NUT

LOCK WASHER

ACCESS AND DRAIN PLUG

VERTICAL BLADE

WASHER

CABLE GUIDE

SCREW

SCREW

BUSHING

UPPER VERTICAL RUDDER

SCREW RIVET

TAIL CONE

HORIZONTAL RUDDER YOKE

HORIZONTAL RUDDER INNER BEARING

SCREW RIVET

SCREW RIVET

HORIZONTAL RUDDER

BUSHING

HORIZONTAL BLADE

SCREW

LOWER VERTICAL RUDDER

HORIZONTAL OUTBOARD BEARING

VERTICAL OUTBOARD BEARING

Figure 2-71. Blades, Rudders, and Linkages, Exploded View.

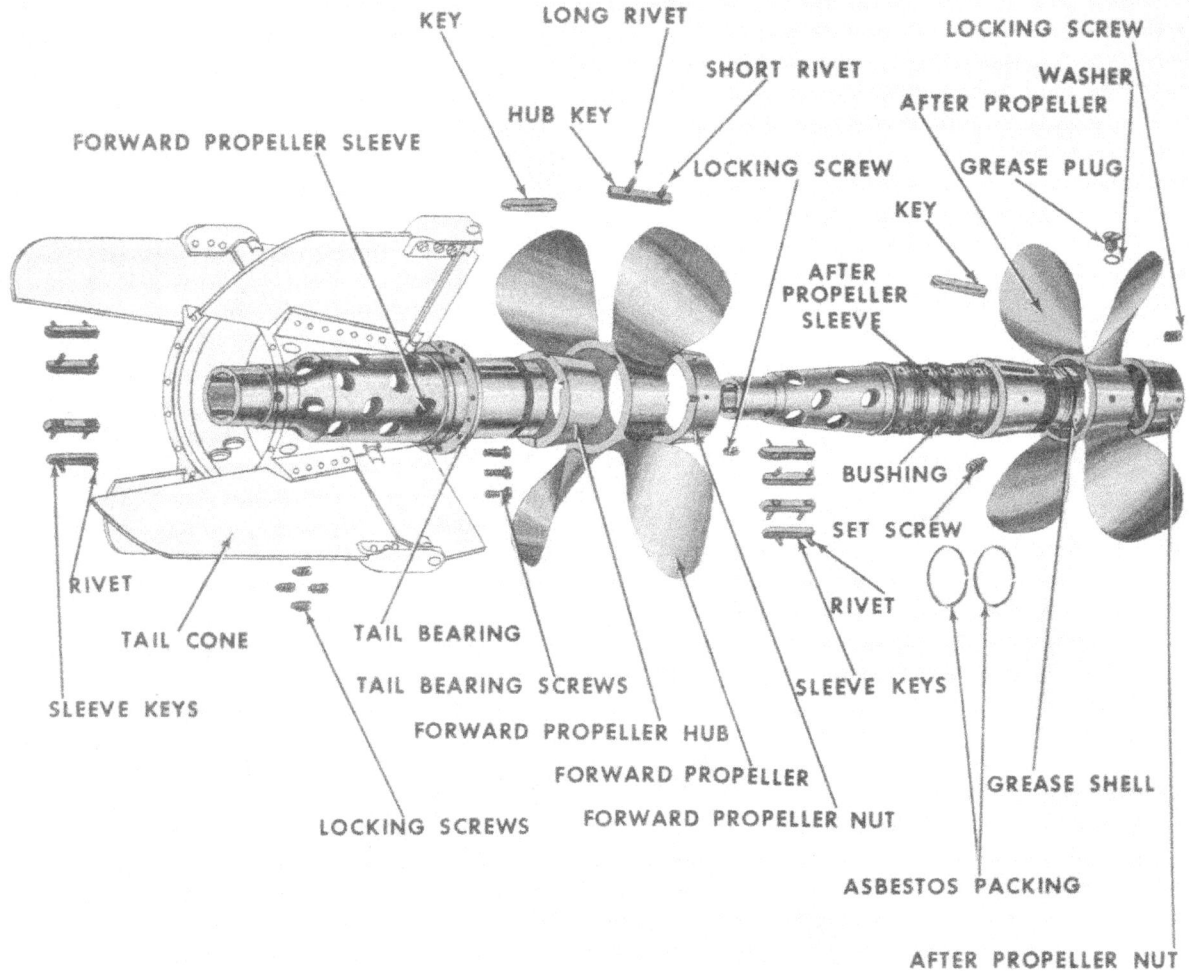

Figure 2-72. Propellers and Mountings, Exploded View.

of the turbine and gear train by four keys riveted to the inside of its forward bore. The after (inner) propeller sleeve is keyed in a similar manner to the after propeller shaft of the turbine and gear train.

2-191. Two methods are combined to lock both propeller sleeves longitudinally in place. The forward end of the forward propeller sleeve is secured to the forward propeller shaft with four setscrews. The forward end of the after propeller sleeve is secured to the after propeller shaft by a retaining nut. Considerable clearance is provided between the tail bearing and the forward propeller sleeve so that expansion and contraction caused by temperature changes will not cause binding at the bearing surfaces.

2-192. The bearing surfaces of the tail assembly are lubricated with grease, which is applied at a counterbored and threaded hole in the after propeller hub and passes through several channels to the tail bearing in the afterend of the tail cone. The order of grease passage is as follows: (1) through

a hole in the after propeller sleeve, which is in alinement with the grease hole in the after propeller hub, to the tubular grease shell installed inside the after propeller sleeve; (2) through holes in the after propeller sleeve to the four grease bushings installed between the two propeller sleeves; and (3) through holes in the forward propeller sleeve to the tail bearing. The tubular grease shell provides a reservoir for the grease. Grooves machined in the four grease bushings distribute the grease evenly for complete lubrication between the bearing surfaces of both propeller sleeves. Grooves machined in the tail bearing distribute the grease evenly for complete lubrication of the tail bearing inner surface and forward propeller sleeve outer bearing surface. When the grease hole in the after propeller hub is not in use, it is sealed with a plug and washer. When the torpedo is running, the hot exhaust gases pass through the port holes in both propeller sleeves and liquefy the grease in the grease reservoir, then centrifugal force provides a continuous supply of liquid grease to the bearing surfaces.

CHAPTER 3

FUNCTIONAL DESCRIPTION

3-1. GENERAL FUNCTIONAL DESCRIPTION OF MAJOR SYSTEMS

3-2. The submarine fire-control system provides three synchro command signals and impulse power to the torpedo (figure 3-1). The synchro commands preset the running pattern in course (azimuth), depth, and enabling run (distance before left circle). The impulse power charges the power pack and fires the torpedo.

3-3. CONTROL SYSTEM. The control system determines the torpedo running pattern from the three synchro command inputs and steers the torpedo along this preset path.

3-4. Course. The gyro-angle order (G) positions a servo, which in turn places the gyro top plate and cam pawls in line with the desired course heading. Deviations from this heading cause the rudder control valve to steer air pressure to one side or the other of the piston in the vertical steering engine. The piston drives the vertical rudders left and right to steer the torpedo in the vertical (azimuth) plane.

3-5. Running Depth. The depth-set order (HQZ) positions a servo, which in turn applies tension to the depth spring of the immersion mechanism. This causes the horizontal steering engine to create a down-rudder condition on the horizontal rudders. The torpedo, upon firing, is subjected to water pressure that is sensed at the diaphragm of the immersion mechanism. When the preset depth is reached, the water pressure and the spring tension balance, and the torpedo levels off.

3-6. Deleted by CHANGE 2

3-7. Fire. The firing pulse explodes a squib in the starting valve to apply pressurized air to both the control system and the propulsion system. The control system uses this air to move the rudders; the propulsion system uses the air to move Navol, fuel, and water through the piping system.

3-8. PROPULSION SYSTEM. The 115-vac input to the power pack charges a group of storage capacitors,

in addition to providing the firing pulse. When the A-cable is severed upon firing, the capacitors discharge to initiate the timing function. The timing circuit opens valves in the proper sequence to permit the Navol, fuel, and water to mix. In the combustion system, the Navol decomposes into hot oxygen and steam. The water sprays into this decomposed Navol to both cool it and provide additional steam. The fuel, aided by the hot oxygen, provides combustion to decompose the Navol and to vaporize the water. The resulting high-velocity hot gases and steam drive the two counter-rotating turbines of the main engine, and the turbines drive the propellers.

3-9. WARHEAD. The warhead contains an exploder mechanism and a search coil with a self-contained electrical system for detonating either on impact or proximity.

3-10. EXERCISE HEAD. The exercise head contains water ballast, which is expelled at the end of the torpedo run to give the torpedo positive buoyancy.

3-11. FUNCTIONAL DESCRIPTION OF CONTROL SYSTEM

3-12. COURSE CONTROL. Course is controlled by the gyro-angle order servo system, the gyro, and the vertical steering engine.

3-13. Gyro-Angle Order Servosystem. The gyro-angle order servosystem (figure 3-2) consists of a computer, a two-speed relay transmitter circuit, an amplifier unit, and a switch box installed in the submarine fire control system; and two synchro transmitters, a servomotor, and gear train installed in the torpedo gyro mechanism.

3-14. When the gyro-angle order (G) is introduced, it is transferred from the computer as a mechanical input to the low-speed, 360° per turn, synchro control transformer (SCT-L) and the high-speed, 10° per turn, synchro control transformer (SCT-H) of the relay transmitter circuit. The low-speed transformer provides a rough control setting. The high-speed transformer provides a close control setting. The output voltage of both transformers represents the amount of error signal.

3-15. The error-signal voltage from the low-speed transformer is fed through the bias unit (BU), the single-speed relay (RY1), and the zero-return relay (RY2) to the transfer circuit (RYT), and the motor amplifier (AM). The error-signal voltage from the high-speed transformer is fed through the single-speed relay, the zero-return relay, the resonant damp (DR), and the low-pass filter (FT) to the transfer circuits, and the motor amplifier. The transfer circuits permit the motor amplifier to pass the

CONTROL SYSTEM

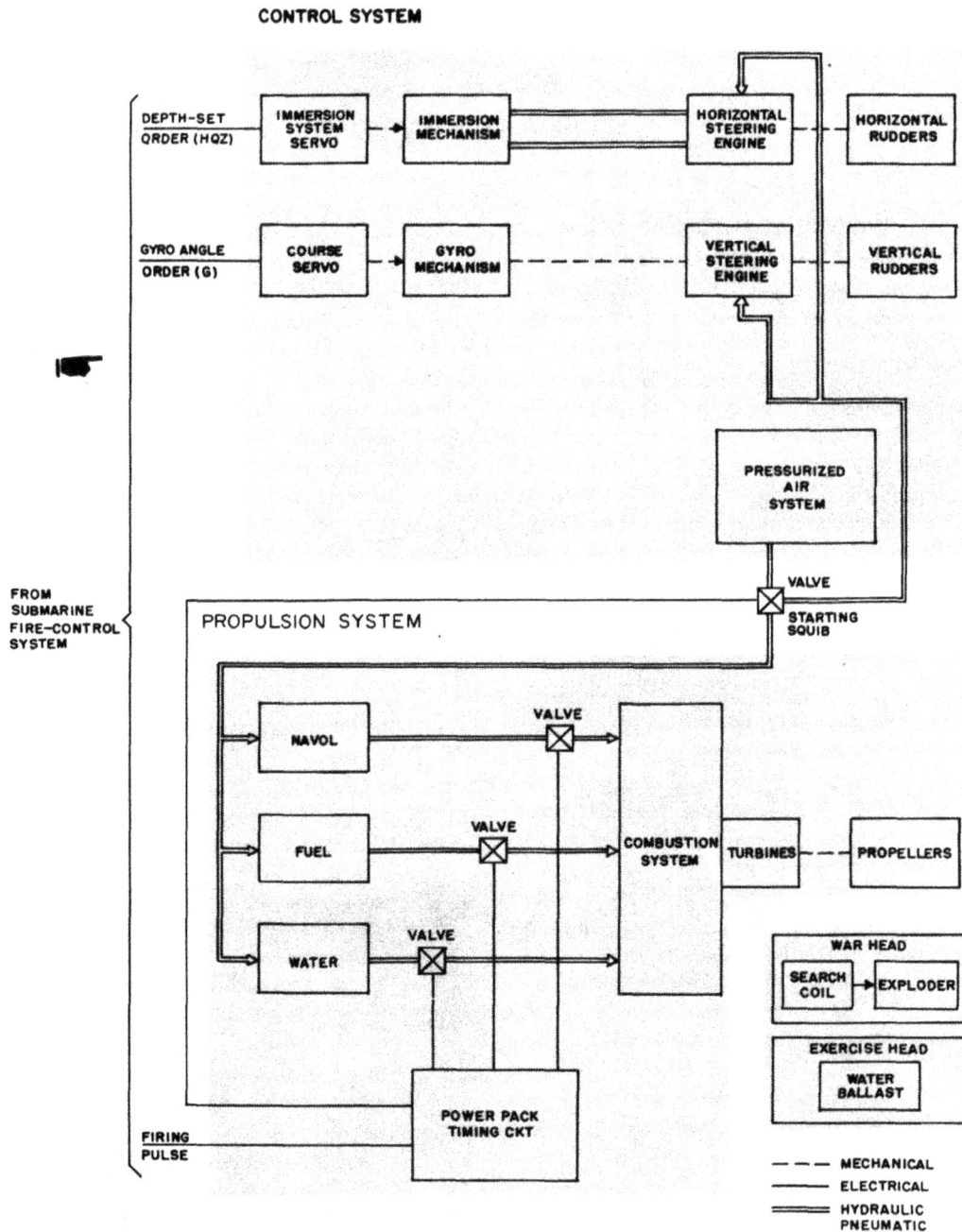

Figure 3-1. Basic Block Diagram.

voltage of the low-speed error signals when the error is large and to pass the voltage of the high-speed error signal when the error is small.

3-16. The amplified error-signal voltage from the motor amplifier drives the servomotor in the torpedo gyro mechanism. The servomotor positions the load to a setting which corresponds with the value of the gyro-angle order transferred from the computer to the synchro control transformers in the fire control system. The servomotor also positions the rotors of the low-speed synchro transmitter and the high-speed synchro transmitter in the torpedo gyro mechanism. The rotors are driven in a direction as required to reduce the error-signal voltage to zero. The servomotor, load, and synchro transmitters are interconnected by a gear train.

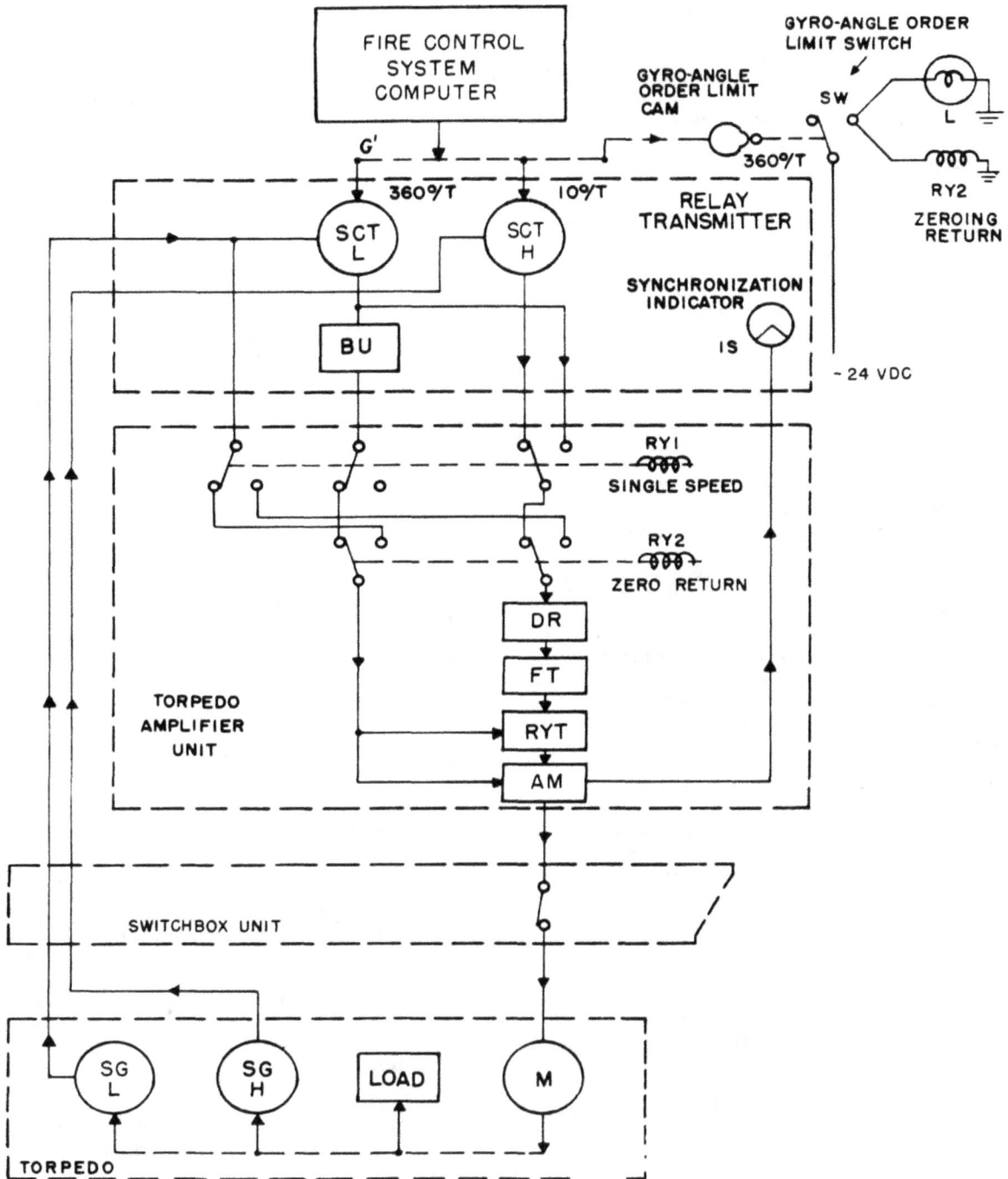

Figure 3-2. Gyro-Angle Order Servosystem.

3-17. A solenoid-type brake mechanism locks the top plate of the torpedo portion of the servosystem to prevent final gyro-angle order setting from being disturbed after torpedo firing. The solenoid-brake mechanism is energized when a gyro-angle order is being automatically set to allow free movement.

3-18. Gyro. The gyro-angle axis of rotation, in relation to a course reference, is held by the centering pin of the spinning and unlocking mechanism before launching. Initial gyro spin is activated from a surge of full-air-flask pressure to the spinning wheel of the unlocking mechanism. When the gyro spin obtains normal speed (12,000 rpm) and unlocks, the full air pressure is cut off and the constant speed is maintained by the working pressure of the gyro reducer valve. Any deviation to port or starboard from the set gyro angle is corrected by the gyro cam pawls and pallet mechanism. Operation of the pallet mechanism is in the form of oscillating mechanical movements of its parts, driven by the pinion gear on the forward propeller shaft.

3-19. When the gyro is spinning at 12,000 rpm and remains freely mounted in its gimbals, it maintains its axis (gyro-angle order setting) regardless of changes in torpedo attitude. The gyro is fixed (locked and spun) on the longitudinal centerline of the torpedo at the start of the run and provides the reference by which the torpedo off-course heading can be corrected throughout the run by the combined operations of the pallet mechanism, vertical steering engine, and vertical rudders (figure 3-3).

3-20. Before torpedo launching, the gyro is set for a torpedo angle shot by altering the position of the pallet mechanism so that it no longer lies directly aft of the vertical centerline of the gyro. The position of the pallet mechanism is altered by the rotation of the gyro-pot top plate. Rotation of the top plate moves the cam pawls and the pallet mechanism away from the gyro cam on the cam plate, which remains fixed with the gyro. Amount of top plate rotation is proportional to the gyro-angle setting. The displacement of the cam pawls and pallet blade of the pallet mechanism causes a full throw of the vertical (course) steering engine, thereby producing a hard-over rudder upon launching. When the torpedo assumes a heading slightly beyond course setting, the engagement of the opposite pawl with the gyro cam causes the pallet blade to turn and, by striking the opposite pawl, to move the connection spool in a vertical plane. Displacement of the connection spool pivots the rock shaft assembly, which causes the vertical steering engine to throw in an opposite direction, thereby changing rudder position. Throughout the remainder of the run, the pallet mechanism makes corrections in course steering as sensed by the gyro when the torpedo heads to one side or the other (port or starboard) from the set course. The resultant oscillating operations of the pallet mechanism, vertical steering engine, and vertical rudders effectively cause the torpedo to slightly zigzag or fishtail on its azimuth course.

3-21. Vertical Steering Engine. The vertical steering engine has a pilot valve and piston operated by the working air pressure. Mechanical linkage connects the pilot valve to the pallet mechanism. Other mechanical linkage connects the piston to the vertical rudders. When the pallet mechanism signals for a starboard rudder position, the pilot valve is moved aft, air pressure enters the aft end of the engine, and the piston is forced forward. As the piston moves forward, residual air pressure is vented from the forward end of the engine body into the afterbody. When the piston reaches its full forward position, the vertical rudders are in a hard-over starboard position. When the pallet mechanism signals for a port rudder position, the pilot valve moves forward,

air pressure enters the forward end of the engine, and the piston is forced aft. As the piston moves aft, air pressure is vented from the aft end of the engine body. When the piston reaches its full aft position, the vertical rudders are in a hard-over port position. The design of the vertical steering engine permits only full-throw conditions of its piston and consequent hard-left and hard-right rudder positions.

3-22. DEPTH CONTROL. Depth is controlled by the depth-set order servosystem, the immersion mechanism, and the horizontal steering engine.

3-23. Depth-Set Order Servosystem. The complete circuit of the depth-set order servosystem (figure 3-4) consists of a computer, a two-speed relay transmitter circuit, an amplifier unit, and a switch box installed in the submarine fire control system and a synchro transmitter, a servomotor, and a gear train installed in the torpedo depth mechanism. Although the fire control system portion of the circuit is equipped for two-speed operation, the torpedo portion of the circuit is equipped for single-speed operation. Therefore, only an explanation of the single-speed operation is given.

3-24. When the depth-set order (HQZ) is introduced, it is transferred from the computer as a mechanical input to the high-speed, 300 feet per turn, synchro-control transformer (SCT-H) of the relay transmitter circuit. The output voltage of the transformer represents the amount of error signal.

3-25. The error-signal voltage from the high-speed transformer is fed through the single-speed relay (RY3), the resonant damp (DR), and the filter (FT) to the transfer circuits (RYT) and the motor amplifier (AM).

3-26. The amplified error-signal voltage from the motor amplifier drives the servomotor (M) in the torpedo depth mechanism. The servomotor positions the load to a setting which corresponds with the value of the depth-set order transferred from the computer to the synchro control transformer in the fire control system. The servomotor also positions the rotor of the synchro transmitter (SG) in the torpedo depth mechanism. The rotor is driven in a direction as required to reduce the error-signal voltage to zero. The servomotor, load, and synchro transmitter are interconnected by a gear train.

3-27. The depth-set order servosystem will be automatically disabled (short circuited) when an input value of the depth-set order exceeds the range in which the torpedo depth mechanism will respond. The disabling (short circuiting) operation is performed by the limit relay (RY4) in conjunction with the depth-set order limit cam and limit switch.

3-28. Immersion Mechanism. Theoretically, the torpedo should approach, level off, and then run at its set depth for the duration of the run. Actually, the torpedo will attempt to overshoot the set depth slightly both up and down before it acquires the set depth. The tendency to overshoot is offset to within a very close approximation of the set depth level by

NOTE
FIRE CONTROL SYSTEMS MK 101, MK 106, MK 112 AND MK 113 PROVIDE ENABLING, DEPTH, GYRO-ANGLE ORDERS, AND STARTING GEAR SIGNAL UP TO TIME OF LAUNCHING.

TORPEDO RUNNING ON DESIRED COURSE. CAM PAWLS NOW IN LINE WITH GYRO AXIS. TORPEDO STRAIGHTENED OUT ON COURSE AND RUDDERS OSCILLATING IN NORMAL MANNER. GYRO AXIS STILL PARALLEL TO ORIGINAL POSITION.

TORPEDO RUNNING ON DESIRED COURSE. RUDDERS OSCILLATING IN NORMAL MANNER.

TORPEDO RUNNING. RUDDERS TURNING TORPEDO TO BRING CAM PAWLS INTO LINE WITH GYRO AXIS. GYRO AXIS STILL PARALLEL TO ORIGINAL POSITION.

TORPEDO READY TO BE FIRED ON STRAIGHT LAUNCH. NO ANGLE SET. CAM PAWLS IN LINE WITH GYRO AXIS.

GYRO
STEERING ENGINE
GYRO TOP PLATE
CAM PAWLS
RUDDER

CAM PAWLS

TORPEDO READY TO BE FIRED ON ANGLE LAUNCH. GYRO TOP PLATE ROTATED TO DESIRED ANGLE, THUS SETTING CAM PAWLS OFF CENTER OF GYRO CAM PLATE BY SAME ANGLE.

STRAIGHT LAUNCH

ANGLE LAUNCH
(Right or left 155° max.)

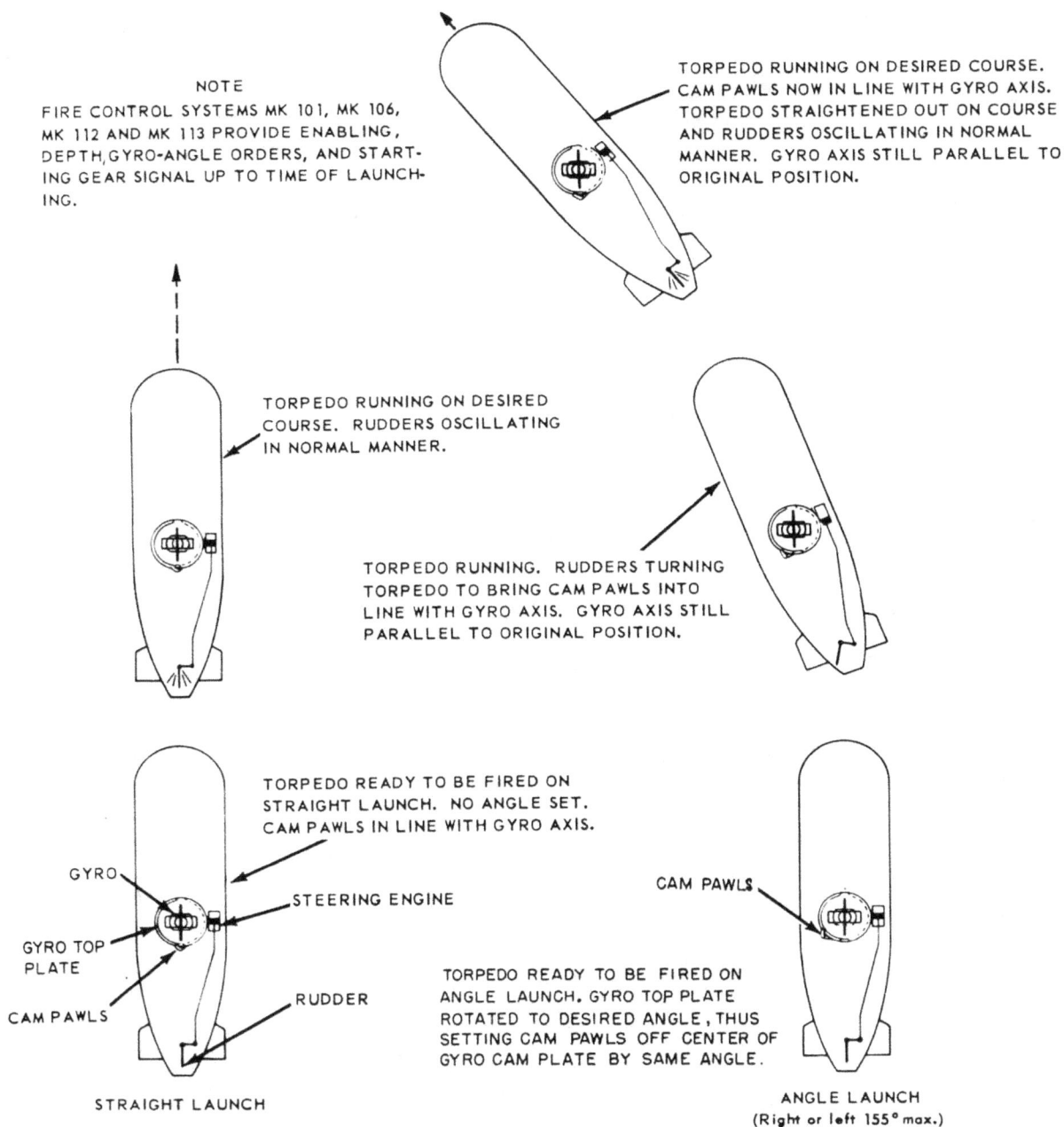

Figure 3-3. Straight and Angle Launchings.

the operation of the immersion mechanism and its associated components. The most undesirable upward overshoot is eliminated by the operation of the climb-limiter device on the pendulum lever of the immersion mechanism. When the torpedo aproaches set depth, the pendulum and spring-loaded hydrodiaphragm of the immersion mechanism work in conjunction to operate the horizontal steering engine which controls the up and down positioning of the horizontal rudders. The entire operation corrects any deviation between actual running depth and set depth. When the set depth is obtained, the entire

operation is continued so that the torpedo will run within ±3 feet of the set depth throughout the remainder of the run.

3-29. The immersion mechanism is the sensitive element of the depth-control system. It is actuated by the downward and upward angular inclination of the torpedo and by the increased and decreased pressures in the variations from set depth. The resulting forces produced by the deviations are applied through linkage to position the valve of the horizontal steering engine. The depth-sensitive portion

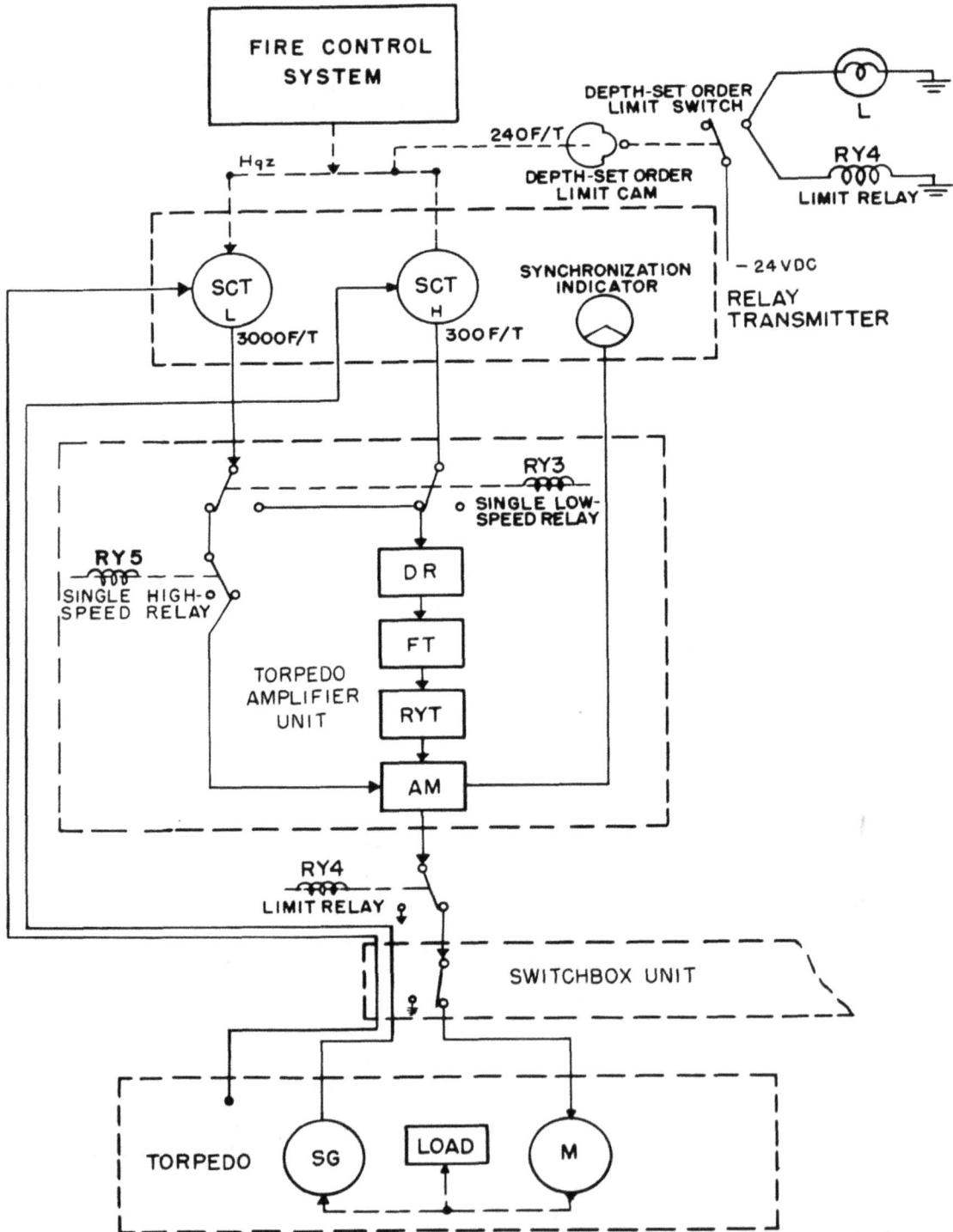

Figure 3-4. Depth-Set Order Servosystem.

of the immersion mechanism is the spring-loaded hydrodiaphragm activated by sea water depth pressure and arranged so that the spring tension can be varied to preset the running depth. When the sea water depth pressure is greater than the preset spring tension, the hydrodiaphragm releases some of the spring tension, thereby causing the valve of the horizontal steering engine to operate as required to obtain an up position of the horizontal rudders. When the sea water depth pressure is less than the spring tension, the entire resulting operations reverse to obtain a down position of the rudders. The angular-sensitive portion of the immersion mechanism is the pendulum which swings forward when the torpedo acquires a nosedown attitude and swings aft when the torpedo acquires a noseup attitude. When the torpedo is nosedown, the pendulum operates the valve of the horizontal steering engine as required to obtain an up position of the horizontal rudders. When the torpedo is noseup, all operations reverse.

3-30. Horizontal Steering Engine. The horizontal steering engine has a valve and piston operated by the working air pressure. Mechanical linkage connects the valve to the pendulum and the spring-loaded hydrodiaphragm. Other mechanical linkage connects the piston to the horizontal rudders. When the torpedo is running deeper than the preset depth order or running in a nosedown attitude, the valve is moved aft by pendulum action, air pressure enters the forward end of the engine body, and the piston is forced aft. As the piston moves aft, the air pressure is vented from the aft end of the engine body into the afterbody. When the piston reaches its full aft position, the horizontal rudders are in a hard-up position. When the torpedo is running too shallow and noseup, the valve moves forward, the air pressure enters the aft end of the piston body, and the piston is forced forward. As the piston moves forward, the air pressure is vented from the forward end of the engine body. When the piston reaches its full forward position, the horizontal rudders are in a hard-down position. However, throughout a normal run, the horizontal steering engine operates with a floating action and can maintain any piston position to set the horizontal rudders anywhere between their up and down extremes.

3-31. Antibroach. At the start of the torpedo run, an overshoot of set depth is controlled by the climb angle-limiter device installed on the pendulum lever of the immersion mechanism. Immediately after the torpedo leaves the tube, the device limits the recovery climb-pitch angle so that the torpedo will not broach and will acquire the set depth quickly and with minimum overshoot. For example, when the torpedo is equipped with the climb angle-limiter device and launched at a 180-foot depth, the torpedo will acquire a low-recovery angle of 10° to 10.5° from horizontal, a minimum or no overshoot will result, and no broach will occur. Otherwise, when the same torpedo is not equipped with the climb angle-limiter device and launched at the same depth, the torpedo will acquire a high-recovery climb-pitch angle of 19° to 21° from horizontal, a great deal of overshoot will result, and a broach may occur.

3-32 thru 3-41. Deleted by CHANGE 2

Figure 3-5. Deleted by CHANGE 2

flask will cause an early water expulsion from the exercise head with the result of an undesirable end-of-run transient.

3-44. Before the torpedo is fully loaded into the tube, the torpedo stop valve (figure 3-6) is opened to allow the full air flask pressure to be banked at the electrically activated starting valve. If the torpedo is to be fired in an exercise shot, the blow valve is also opened to allow the air flask pressure to be banked at the air-release mechanism in the exercise head. When the torpedo is launched, the starting valve is electrically activated (squib-fired) by the submarine fire control system to allow the air flask pressure to flow to the spinning and unlocking mechanism to give the gyro its initial spin and to the pressure reducer to be reduced to a working air pressure of 635 ±10 psi. The full air flask pressure for initial gyro spin is fed through the gyro spinning mechanism and serves only to bring the gyro up to running speed as quickly as possible. Within 0.3 second, the correct gyro speed is obtained and the gyro spinning mechanism becomes inoperative. An impulse valve in the gyro spinning mechanism automatically cuts off the flow of air flask pressure. Gyro spin is maintained at a constant speed throughout the run by the 125 psi air pressure from the gyro reducing valve.

3-45. Upon its delivery from the pressure regulator, the working air pressure (635 ±10 psi) is used for these purposes: (1) to pressurize the Navol tank, water compartment, and fuel tank for subsequent delivery of the liquids to the combustion system at start-of-run; (2) to operate the horizontal and vertical steering engines and gyro reducing valve of the control mechanism throughout the run.

3-42. ENERGY CONTROL. The energy section stores air, water, fuel (alcohol), and Navol, from which energy is produced; it also provides devices for controlling energy from start to finish of a torpedo run. The air and fluids are delivered to the afterbody in correct amounts and sequence for use in producing propelling forces. The devices that control the flow of air and fluids consist of a valve system and an electrical power system. At the end of a torpedo exercise run, a portion of the valve and power systems controls the flow of air pressure to expel water ballast from the exercise head and to expel remaining fuel and water from the energy section to give the torpedo positive buoyancy. All Navol is expended upon completion of a normal run. However, some Navol remains in Navol tank after end of an interrupted run.

3-43. Air Pressure Distribution. Before firing a torpedo war shot, the air flask must be charged to a pressure of 2800 ±50 psi and not allowed to decrease to less than 2750 psi to insure a full run. Before firing an exercise shot, it is recommended that the air flask be charged as close as possible to 2800 ±50 psi. Otherwise, an overcharged air flask will cause a delayed water expulsion and an undercharged

3-46. At the start of run, the working air pressure passes through the manifold and the air-check portion of each Navol, water, and fuel air check and vent valve before it reaches the liquid compartments. However, before entering the Navol tank, it passes through the rupture disk and screen assembly. After the liquid compartments are pressurized, the liquids are delivered to the combustion system via the Navol, water, and fuel delivery valves which are electrically activated in proper sequence by the operation of the power pack.

3-47. Throughout the run, the working air pressure operates the horizontal and vertical steering engines, and gyro reducing valve as follows: At the horizontal steering engine, working air pressure is ported to the correct side of the engine piston by the operation of the engine pilot valve which is controlled by the immersion mechanism. At the vertical steering engine, working air pressure is ported to the correct side of the engine piston by the engine pilot valve which is controlled by the rudder-control valve and gyro mechanism. At the gyro reducing valve, it is reduced to 125 psi to maintain a constant gyro-spinning speed.

3-48. Deleted by CHANGE 2

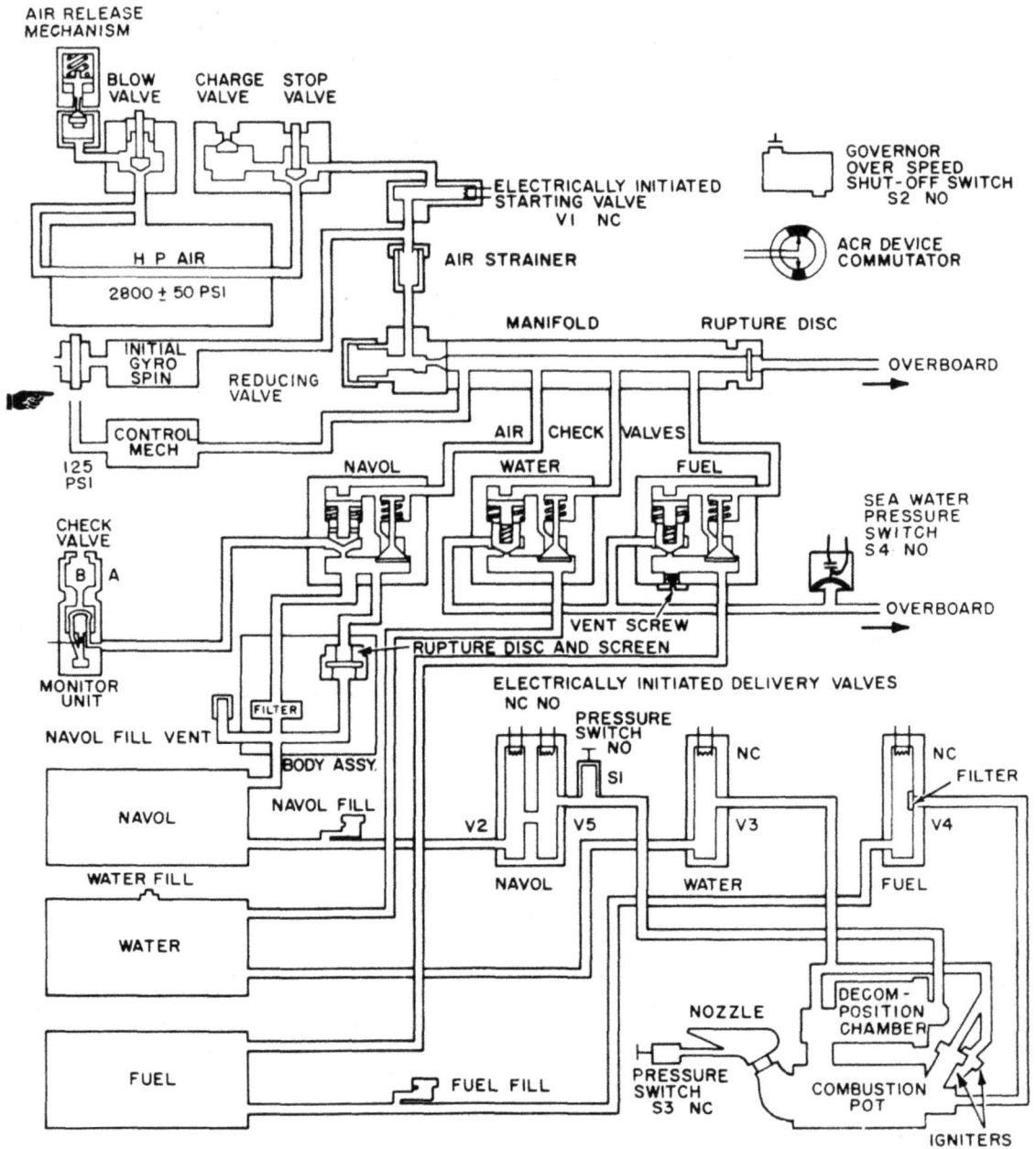

Figure 3-6. Pneumatic and Hydraulic System.

Deleted by CHANGE 2

release mechanism in the exercise head opens, and the decreased air flask pressure enters the exercise head and forces the water ballast out through the two expulsion valves into the sea. In addition, the remaining air flask pressure is used to expel most of the residual water and fuel overboard from their compartments through the exhaust system and vent manifold and most of the residual Navol overboard from the Navol tank through the Navol vent valve and monitor unit check valve to provide extra buoyancy.

3-49. At the end of an exercise run, when the air flask pressure decreases to 1450 ±25 psi, the air-

3-50. Timing for Fluid Distribution. Start-of-run is initiated when 115-volt 60-Hz impulse power is applied by the submarine fire control system. This impulse power passes through Switchbox Mk 5 or Mk 11, external torpedo (B- then A-) cabling, and internal torpedo cabling, reaching the power pack through sea-actuated switch S4. After impulse power is applied, squibs of electrically activated starting (air) valve V1, Navol delivery valve V2, water delivery valve V3, and fuel delivery valve V4 of the pneumatic and hydraulic system (figure 3-6) are fired in correct sequence by the power pack (figure 3-7). After Navol delivery valve V2 is actuated, switch S1 of the energy control system is closed by the pressure of Navol delivery. Although the gyro of the anti-circling-run (ACR) device is not directly involved with start-of-run, it is electrically (squib) actuated during the start-of-run sequence of power pack operation.

3-51. The power pack is a simple timing circuit (figure 3-7) made up of four time delay relays (K1, K2, K3, and K4), four diodes (CR1, CR2, CR3, and CR4), four capacitors (C1, C2, C3, and C4), and four resistors (R1, R2, R3, and R10). Another time delay relay (K1) is mounted remote from the power pack and external to the ACR device. The timing sequence of the circuit is controlled by relay K3 with a delay on pull-in of 100 milliseconds, relay K1 with a delay on dropout of 50 milliseconds, relay K2 with a delay on dropout of 360 milliseconds, K4 with a delay on dropout of 10 seconds, and K1 (ACR) with a delay on dropout of 55 seconds. All relays are energized at the same time; thus, the 100-millisecond delay on pull-in of relay K3 allows the capacitors ample time to become fully charged through the normally closed contacts of the relays and the diodes and resistors before the firing impulse power is lost upon the cutting of the torpedo A-cable. Delay-on-dropout relays K1, K2, K4, and K1 (ACR) provide the accurate control of the timing sequence for discharging the stored energy from the capacitors to fire the squibs of the electrically actuated components of the energy control system and ACR.

3-52. At the power pack, the firing-impulse power enters through pins A and B of connector 2A3P6 to energize relay K3 and pass through normally closed contacts 2 and 3 of relay K3 to energize relays K1, K2, K4, and K1 (ACR). Contacts 2 and 3 of relay K3 remain closed during the 100-millisecond delay on pull-in of relay K3. At the instant relay K3 is energized, the firing-impulse power passes through contacts 2 and 3 of relay K3, contacts 6 and 7 of relay K2, and current-limiting resistor R1 to fire the squib of the starting (air) valve V1. At the same instant, the firing-impulse power passes through contacts 2 and 3 of relay K3, contacts 6 and 7 of relay K2, contacts 6 and 7 of relay K4, and current-limiting resistor R2 to fire the squib of the Navol delivery valve V2. Also, at the same instant, the firing impulse is passed through contacts 2 and 3 of relay K3, contacts 6 and 7 of relay K2, contacts 6 and 7 of relay K1 (ACR), and current-limiting resistor R10 to fire the squib used in starting ACR gyro.

3-53. During the 100-millisecond delay on pull-in of relay K3, capacitors C1, C2, C3, and C4 are charged simultaneously. Firing-impulse power passes through current-limiting resistor R3, contacts 2 and 3 of relay K1, and diode CR1 to charge capacitor C1 with half-wave pulses. In the same manner, firing-impulse power passes through current-limiting resistor R3 for charging of the remaining capacitors C2, C3, and C4, but with the following routine changes: (1) contacts 2 and 3 of relay K2 and diode CR2 to charge capacitor C2 with half-wave pulses; (2) contacts 6 and 7 of relay K1 and diode CR3 to charge capacitor C3 with half-wave pulses; and (3) contacts 6 and 7 of K1, contacts 2 and 3 of K4, and diode CR4 to charge capacitor C4 with half-wave pulses.

3-54. When the 100-millisecond delay on pull-in of relay K3 terminates, contacts 2 and 3 of relay K3 open to deenergize delay-on-dropout relays K1, K2, and K4. Then, 50 milliseconds later, or 150 milliseconds after the start of run, contacts 2 and 4 of relay K1 close and capacitor C1 discharges to fire the squib of water-delivery valve V3. At the same time, contacts 6 and 7 of relay K1 open to prevent capacitors C3 and C4 from being discharged through the back resistance of the diodes, relay coils, and possibly short-circuited A-cable. Thus, capacitors C3 and C4 remain charged until acted on by the ACR gyro, governor switch S2, and pot pressure sensing switch S3, as explained later in this chapter under the descriptions of the methods of torpedo shutdowns.

3-55. Approximately 400 milliseconds after firing-impulse power is applied, pressure builds up in the Navol delivery pipe as required to actuate (close) switch S1, thereby completing the circuit between open contact 4 of relay K2 and the squib of fuel-delivery valve V4. Then, 460 milliseconds after firing-impulse power is applied, which is a total of the 100 milliseconds delay on pull-in of relay K3 plus the 360 milliseconds delay on dropout of relay K2, contacts 2 and 4 of relay K2 close and capacitor C2 discharges through contacts 2 and 4 of relay K2 and switch S1 to fire the squib of the fuel-delivery valve V4.

3-56. Provided the power pack operated correctly, the following sequence occurs. When the squib of the starting (air) valve V1 is fired, the valve opens to deliver air-flask pressure (2800 psi nominal) to give Gyro Mk 12 Mod 3 its initial spin and through the air strainer to the pressure reducer to be reduced to the working air pressure. When the squib of Navol delivery valve V2 is fired, the valve opens to deliver Navol through the normally open Navol shutdown valve V5 to the Navol restriction inlet of the decomposition chamber. When the squib of water-delivery valve V3 is fired, the valve opens to deliver water to the water-spray inlet of the decomposition chamber and to Igniters Mk 6 Mod 4 of the combustion pot. The igniters are fired by the pressure of the water. When the squib of fuel-delivery valve V4 is fired, the valve opens to deliver the fuel to the fuel-spray inlet of the combustion pot.

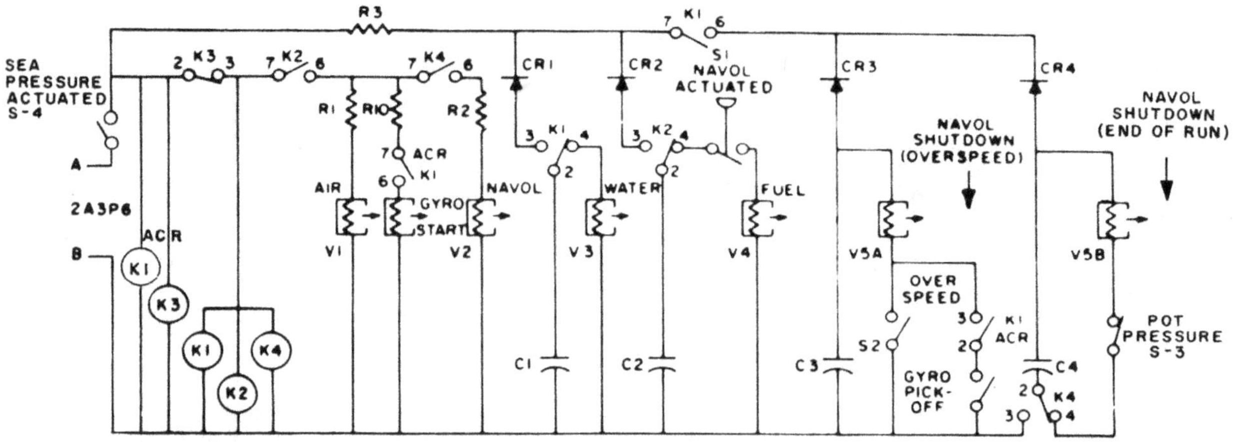

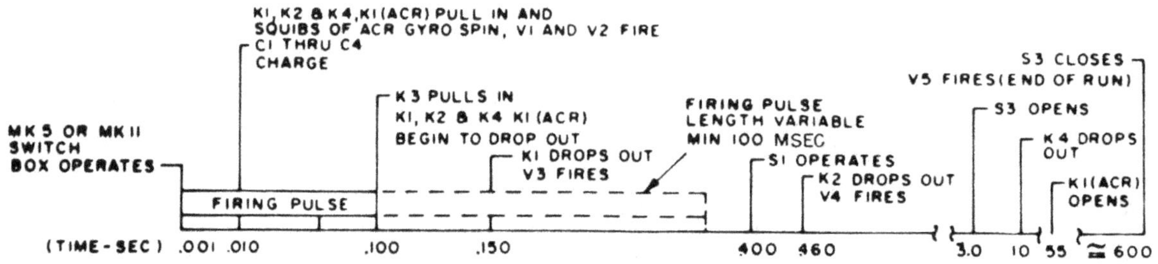

Figure 3-7. Diagram and Sequential Time Chart of Power Pack.

NOTE:

(1) BEFORE TORPEDO CAN BE FIRED, TUBE MUST BE AT A MINIMUM DEPTH OF 10 FEET SO THAT SUFFICIENT SEA PRESSURE IS AVAILABLE TO CLOSE SWITCH S4.

(2) WHEN FIRING IMPULSE VOLTAGE IS APPLIED, SQUIB-ACTUATED AIR DELIVERY VALVE V1 AND NAVOL DELIVERY VALVE V2 OPEN; NAVOL DELIVERY PRESSURE CLOSES SWITCH S1; CAPACITORS ARE CHARGED; TIME DELAY RELAYS ARE ENERGIZED; AND SQUIB-ACTUATED WATER DELIVERY VALVE V3 AND FUEL DELIVERY VALVE V4 OPEN.

(3) IF TORPEDO MAIN ENGINE OVERSPEEDS AT ANY TIME AFTER FIRING, SWITCH S2 CLOSES TO INITIATE NAVOL DELIVERY SHUTDOWN FOR END OF RUN.

(4) WHEN COMBUSTION POT PRESSURE REACHES 400 PSI WITHIN 10 SECONDS AFTER FIRING, SWITCH S3 OPENS AND REMAINS OPEN THROUGHOUT A NORMAL RUN. IF SWITCH S3 DOES NOT OPEN BECAUSE OF IGNITION FAILURE OR OTHER REASON AND WHEN IT RECLOSES AT END OF NORMAL OR INTERRUPTED RUN, IT INITIATES NAVOL DELIVERY SHUTDOWN FOR END OF RUN.

(5) IF TORPEDO ATTAINS A COURSE HEADING OF 165° TO LEFT OR RIGHT OF 0° GYRO-ANGLE SETTING WITHIN 55 SECONDS AFTER FIRING, PICKOFF SWITCH OF ACR GYRO CLOSES TO INITIATE NAVOL DELIVERY SHUTDOWN FOR END OF RUN.

3-57. A successful start of torpedo run is made when the delivery valves operate correctly, water pressure fires the igniters, and combustion pot pressure builds up sufficiently to open pot pressure sensing switch S3. The torpedo will make a complete run if the 10-second delay-on-dropout of relay K4 closes its contacts 2 and 4 while pot pressure sensing valve S3 remains open, if the main engine does not overspeed to cause governor switch S2 to close, and

if the 55-second delay-on-dropout relay K1 opens its contacts 2 and 3 before the gyro pickoff switch closes upon operation of the anti-circling-run (ACR) device.

3-58. FUNCTIONAL DESCRIPTION OF PROPULSION SYSTEM

3-59. COMBUSTION SYSTEM. Operation of the combustion system (figure 3-8) begins during the initial

DECOMPOSITION CHAMBER
POT PRESSURE SENSING VALVE NIPPLE
SPACER
HYDROGEN PEROXIDE DECOMPOSED TO OXYGEN AND WATER
WATER PRESSURE
CATALYST CARTRIDGE
NAVOL
NOZZLE
WATER
IGNITERS
FIRING IGNITERS
NOZZLE GASES
FUEL
COMBUSTION POT

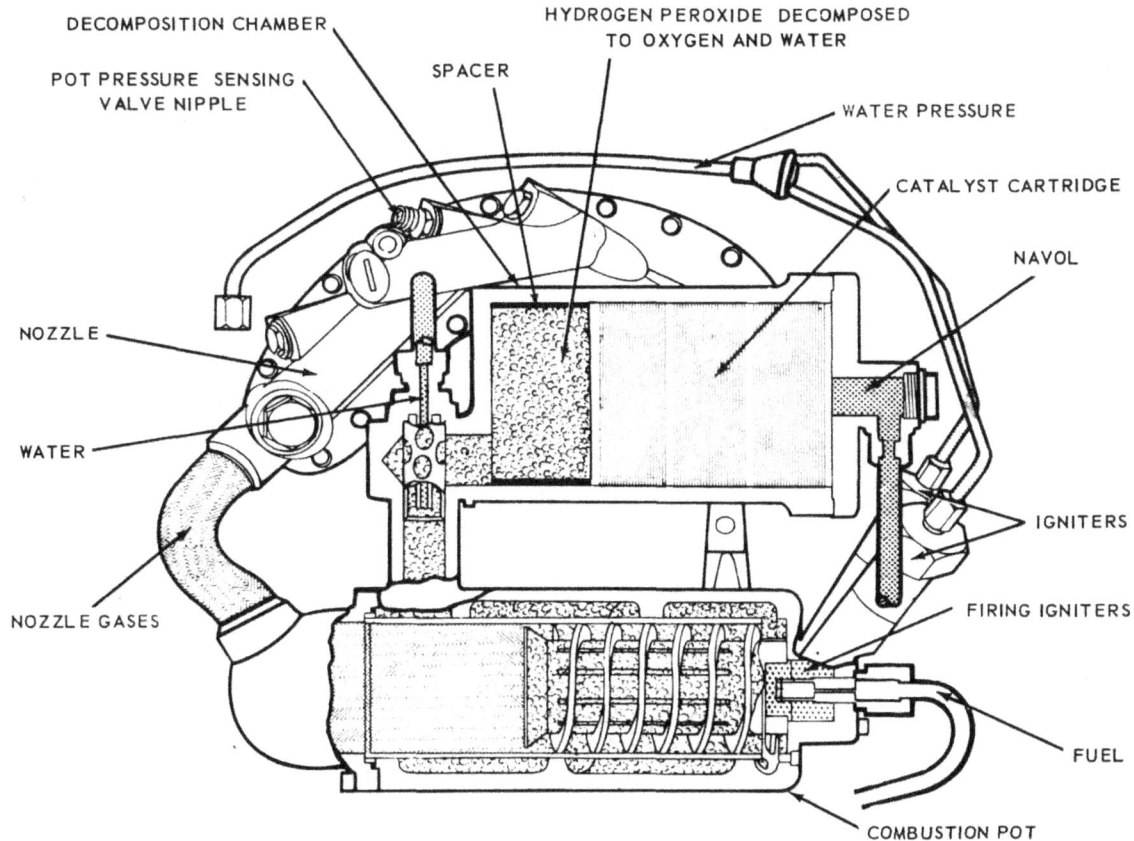

Figure 3-8. Combustion System.

portion of the start of the torpedo run and depends on the proper sequence of Navol, water, and fuel delivery and igniter actuation. The proper sequence is controlled by the power pack and valve system in the torpedo energy section. Because the sequence is so rapid, the combustion system produces 85 percent of its operational efficiency in 700 milliseconds of the start of torpedo run. Within this length of time, the torpedo will be completely ejected from the tube and have enough power to assume its own propulsion before the combustion system operation obtains peak efficiency.

3-60. Approximately 100 milliseconds after the start of torpedo run, the Navol is delivered through a restriction and enters the decomposition chamber of the combustion system. In the decomposition chamber, the Navol is forced through a series of catalyst screens to be decomposed into hot oxygen and steam which pass through a hollow stud and enter the combustion pot of the combustion system. When the hot oxygen and steam pass through the hollow stud, they combine with the water delivered through a restriction and spray nozzle on the hollow stud in approximately 150 milliseconds after the start of torpedo run. At the same time, the water pressure is delivered to the igniters for their actuation. Then, in approximately 460 milliseconds after the start of torpedo run, the fuel (alcohol) is delivered through a

restriction and spray, enters the combustion pot, and is set on fire by the igniters. In the meantime, the combined hot oxygen, steam, and water are thoroughly mixed as they pass through the inner and outer whirls in the combustion pot before joining with the ignited fuel.

3-61. The hot oxygen supports the combustion of fuel. The water keeps the combustion temperature at a safe level and becomes additional steam. The two igniters burn for approximately 5 seconds after they are actuated to doubly insure the initial firing of the fuel. Throughout the remainder of the run, the burning of the fuel is self-sustaining with the product of combustion consisting of high-velocity gases and steam expelled under pressure from the combustion pot through four oblique ports of the nozzle. These gases are directed on the buckets of the turbines of the main engine. On leaving the buckets, the hot gases and steam pass through the torpedo exhaust system into the sea. A quite inconspicuous torpedo wake exists because the steam is condensed into water vapor and most of the gases are absorbed in the sea.

3-62. TURBINE AND GEAR TRAIN. The turbine and gear train (figure 3-9) uses high-velocity hot gases and steam delivered from the nozzle of the combustion system, transferring the kinetic energy

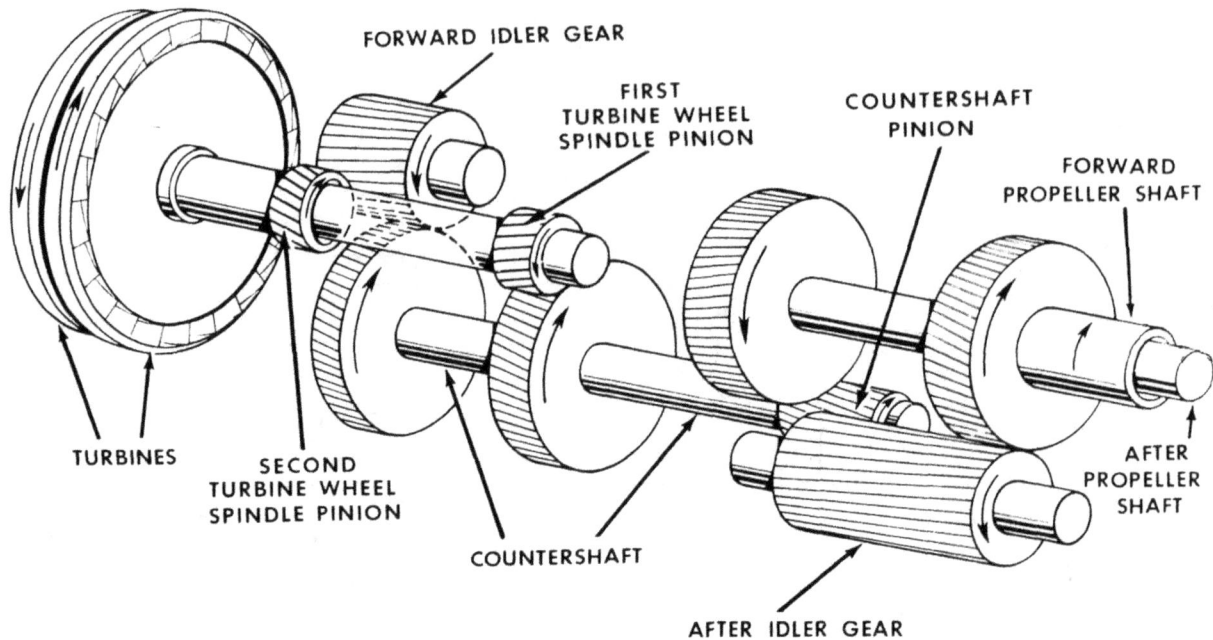

Figure 3-9. Turbine and Gear Train.

of the hot gases and steam into mechanical power to drive the torpedo propellers. The basic means of propulsion is obtained from two counterrotating turbine wheels, a gear train with a 9 to 1 ratio of reduction, and two counterrotating propeller shafts mounted in the engine frame. Because the torpedo propellers are mounted on sleeves connected to the shafts, they also rotate in opposite directions. When the turbine and gear train is operating at full speed and under a normal load, both turbine wheels rotate at 12,870 rpm and both propellers rotate at 1430 rpm because the gear train has a 9 to 1 ratio of reduction. A balancing effect is produced by the combined operation of the gear train and opposite forces applied to the turbine wheels and against the propellers.

3-63. When the hot gases and steam are expelled from the oblique ports of the combustion system nozzle, they are directed toward the curved buckets of the forward (or first) turbine wheel and cause the wheel to spin in a counterclockwise direction (looking forward). On leaving the buckets of the first turbine wheel, the hot gases and steam are directed toward the oppositely curved buckets of the second turbine wheel, causing the wheel to spin in a clockwise direction. On the first turbine wheel spindle is a pinion gear that drives a mating gear on the countershaft. A pinion gear on the second turbine wheel spindle drives the forward idler gear that drives another gear on the countershaft. This portion of the gear-train ratio balances the speed of the two turbine wheels and drives the countershaft in a clockwise direction. On the countershaft is a pinion gear that drives the gear on the after propeller shaft and also the after idler gear that drives the gear on the

forward propeller shaft. This portion of the gear-train ratio balances the speed and causes the opposite rotation of the forward and after propeller shafts.

3-64. As a result of turbine and gear-train operation, the forward propeller rotates in a clockwise direction (looking forward), while the after propeller rotates in a counterclockwise direction. A steady and balanced forward thrust for torpedo propulsion is obtained from the proper pitch and curvature of the propeller blades with respect to the counterrotating motion of the two propellers.

3-65. FUNCTIONAL DESCRIPTION OF WARHEAD

3-66. Warhead Mk 16 Mod 7 operates independently of the remaining torpedo sections; but detonation of its main explosive charge HBX-3 is controlled by Exploder Mk 9 Mod 7 and Search Coil Mk 28 Mod 0 operating in conjunction with one another, or by the exploder alone. When the torpedo scores a direct hit on the enemy vessel, the exploder contact firing mechanism alone can initiate detonation of the warhead main explosive charge. When the torpedo is in the proximity of the enemy vessel, the combined influence-firing circuits of the exploder and search coil detonate the warhead main explosive charge.

3-67. The exploder is primarily intended for combined contact and influence firing, although it may be used in contact-only firing. It is not recommended that the exploder be used for contact-only firing except in a torpedo launching in shallow water or other tactical situations that may necessitate this type of usage. Before the exploder can be used for contact-only firing, the influence-firing circuits must be deactivated by an external means before loading the torpedo in the tube.

3-68. There is a great advantage when the exploder is used in combined contact and influence firing because this detonates the main explosive even though the torpedo may not strike the target. If the torpedo does not make a direct hit but passes close under the keel of the enemy vessel (because of an overestimated depth of enemy vessel keel or erratic torpedo depth performance), the influence-firing circuits will still cause main explosive detonation with more damage resulting than if the torpedo made a direct hit. In a contact-only firing, the running depth would have to be reduced to compensate for an overestimated keel depth or erratic depth performance. With combined contact and influence firing, greater running depths can be set, and there is a greater chance of destroying the enemy vessel.

3-69. For example, assume an enemy vessel keel is at an estimated depth of 10 feet. With the contact-only firing, it would be risky to use the minimum depth setting of 10 feet because actual keel depth may be 2 or 3 feet less than the estimated depth and actual torpedo running depth may be 2 to 3 feet deeper than the set depth. With influence firing, the running depth can be set at 10 feet without risk because the torpedo will pass within 4 feet of the enemy keel even when the keel depth is actually 8 feet and torpedo running depth is actually 12 feet. When the torpedo passes within 4 feet of the enemy vessel keel, detonation by influence firing is certain even though the enemy vessel has a very narrow beam.

3-70. In addition, the torpedo makes a better run at a deeper depth because of less surface wave interference; it is less likely to be sighted by the enemy, and it inflicts more damage. Therefore, influence firing makes it possible to complete successful attacks on enemy vessels with a keel depth too shallow to make contact-only firing feasible. It is recommended that torpedo running depth be set as close as possible to the estimated enemy vessel keel depth so that a low-contact or a proximity hit will occur. In either case, damage at or near keel depth will exceed the damage inflicted by a more shallow contact hit. In the case of a large enemy vessel, the damage would be inflicted below the protecting armor. When the torpedo becomes expended, misses its target, and sinks, a floor switch in the exploder prevents detonation of the main explosive charge when the torpedo hits the sea bottom, or when the torpedo reaches a depth where pressure is great enough to crush the afterbody. (The reaction of the afterbody enabling could cause the exploder inertia switch to operate.)

3-71. Refer to OP 3370 for detailed operational descriptions of Exploder Mk 9 Mod 7.

3-72. FUNCTIONAL DESCRIPTION OF EXERCISE HEAD

3-73. Exercise Head Mk 84 Mod 0 facilitates torpedo recovery by giving the torpedo the necessary positive buoyancy for end-of-run recovery. The exercise head contains water ballast, Air Release Mechanism Mk 3 Mod 5, and expulsion valves to provide buoyancy at the normal end-of-run; a Depth and Roll Recorder (D&R) Mk 2 Mod 1 (optional) to record torpedo attitude during the run; and a Locating Device (Pinger) to locate a sunken torpedo when buoyancy fails to be provided. A dye may be added to the water ballast to increase the probability of sighting a floating torpedo.

3-74. WATER BALLAST EXPULSION. The water ballast expulsion (or buoyancy) function of the exercise head is initiated by the release of the residual air-flask pressure through the air-release mechanism. Before the torpedo is loaded into the tube, the stop plug is removed from the blow valve installed in the forward end of the energy section. Upon removal of the stop plug, air-flask pressure cocks and holds the air-release mechanism in a closed position. The air-release mechanism remains closed until its preset spring pressure overcomes the diminished pressure remaining in the air flask at the normal end-of-run. When the spring tension becomes greater than the residual air-flask pressure, the valve opens to admit air pressure into the exercise head. When the interior head pressure is great enough, the spring-loaded expulsion (discharge) valves are forced open and the water ballast is expelled through the valves into the sea. When the air-flask pressure decreases to the point where the spring tension on the expulsion valves is greater, the valves close to exclude sea water entry and trap air in the exercise head for torpedo buoyancy.

3-75. DEPTH AND ROLL RECORDING. The depth and roll of a torpedo proofing or Fleet exercise run is recorded by Depth and Roll Recorder (D&R) Mk 2 Mod 1. This recorder is a self-contained, mechanically operated mechanism installed in an exercise head flange hole, if used, and manually armed before loading the torpedo into the tube. It produces a permanent calibrated record of the actual change in depth and roll throughout the run. Refer to OP 1711 for detailed operational descriptions of the depth and roll recorder.

3-76. DETAILED FUNCTIONAL DESCRIPTION OF CONTROL SYSTEM

3-77. COURSE CONTROL. The gyro, in conjunction with other components, maintains course control. The gyro setter assembly sets the course to the desired heading; the spinning and unlocking mechanism locks the gyro prior to a torpedo run, causes the initial gyro spin, then unlocks the gyro after it reaches its normal spinning speed; the pallet mechanism transmits gyro error to the vertical steering engine; and the vertical steering engine corrects for course deviation sensed by the gyro.

3-78. Gyro-Setter Assembly. Synchro transmitter SG-L (figure 3-2) rotates at a low (single) speed rate of 360° per turn. Synchro transmitter SG-H rotates at a high (times-36) speed rate of 10° per turn. When synchro transmitter SG-L is operating with the fire control system, its stator windings are inter-connected with the stator windings of low-speed (360° per turn) synchro control transformer SCT-L in the relay transmitter of the fire control system. When synchro transmitter SG-H is operating with the fire control system, its stator windings are interconnected with the stator windings of high-speed (10° per turn) synchro control transformer SCT-H in the relay transmitter of the fire control system. A reference voltage to energize the rotors of both synchro trans-formers is obtained from one phase of the firing craft 115-v 400-Hz three-phase power source.

3-79. Servomotor M1, a two-phase induction motor, drives the gear train. Its rotor turns 144 revolutions for each revolution of the rotor of synchro transmitter SG-L. Its main field is energized by a constant am-plitude from one phase of the firing craft 115-v 400-Hz three-phase power source. Its control field is energized by a variable amplitude of the 400-Hz volt-age, which is the amplified error voltage from synchro control transformers SCT-L and SCT-H in the relay transmitter of the fire control system. The two field voltages must have a phase difference of 90°. Resistor R1 and capacitor C1 of the R-C network are shunted across the control field of servo-motor M1 to provide additional phase shift to the control field voltage and to insure a quadrature with the main field voltage.

3-80. Upon being driven by servomotor M1, the gear train drives the gear on the gyro top plate. When synchro control transformers SCT-L and SCT-H receive a gyro angle order from the fire control system, they cause servomotor M1 to drive the gear train in the direction required to set the gyro top plate. A flywheel on the servomotor shaft keeps the natural frequency of vibration within suitable limits. The large radial ball bearing mounting between the gyro top plate and the gyro pot, plus the large gear ratio between the servomotor shaft gear and the gyro top plate gear, allow the small torque of the servo-motor to position the load with a minimum of friction error. The driving gear that engages the gyro top plate gear is under and near the left-hand side of the mounting bracket.

3-81. Gyro-setter brake solenoid S1 consists of a solenoid, brake arm, brake disk, and compression spring mounted between the brake arm and solenoid core. The brake disk is mounted on top and near the center of the mounting bracket. When the brake solenoid is not energized, small teeth on the brake arm engage similar teeth on the brake disk rim and the spring forces the brake arm inward, pushing the end of the brake arm against the brake disk to lock the gear train. When the brake solenoid is energized, the solenoid core pulls the brake arm outward against the spring compression and the brake disk allows the gear train to operate.

3-82. Gyro. The gyro is fixed (locked and spun) on the longitudinal centerline of the torpedo at the start of the run and provides the reference by which the torpedo off-course heading can be corrected during the run by the combined operation of the pallet mech-

anism, vertical steering engine, and vertical rudders. The gyro maintains its axis throughout the run, re-gardless of torpedo attitude. Refer to OP 627 for a detailed operational description of the gyro.

3-83. Spinning and Unlocking Mechanism. The lock-ing and unlocking gear train (figure 3-10) does the following: (1) engages and disengages the centering pin at the locking end bearing of the gyro inner gimbal ring, (2) engages and disengages spinning gear with gyro wheel gear, and (3) opens and closes the im-pulse valve to control the air flask pressure supply to the spinning wheel. Engaging the centering pin and spinning gear plus opening the impulse valve are done manually when the gyro is installed in the gyro pot and the spinning and unlocking mechanism is locked. Disengaging the centering pin and spinning gear plus closing the impulse valve take place automatically at the start of the torpedo run when the spinning and unlocking mechanism unlocks.

3-84. When the gyro is installed in the gyro pot, the duration-of-spin (unlocking time) adjustment is cocked using an L-shaped tool. A locking lever extends through a slot in the forward face of the frame and is keyed to the shaft on which the spring-loaded bell crank is mounted. A socket in the locking lever re-ceives the shorter arm of the cocking tool. When the locking lever is pulled back, the spring-loaded bell crank moves in a forward position and its spring is compressed. At the same time, the swivel block on the valve bell crank shaft moves up to unseat the impulse valve. A pinion gear, mounted at top of the two bell cranks, drives the upper rack, spinning wheel, spinning gear, and their connecting shaft to a forward position. At the start of forward travel, the toe at the after end of the upper rack is drawn through a slot in the lower rack and passes over the smaller diameter of the centering pin. Midway in forward travel, the upper rack toe contacts a shoulder in the centering pin to start forward motion of the centering pin. When the gyro is being locked, it must be manipulated at the same time the spinning and unlocking mechanism is being cocked so that the centering pin and spinning gear are correctly en-gaged with the gyro. It is usually necessary to rotate the spinning gear slightly to mesh with the gear on the gyro wheel. To make installation easier, the gears are chamfered and the centering pin has a ball-shaped end. At the end of forward travel, a spring-loaded hand-trip lever is forced to the left by the upper rack and snaps over the afterend of a locking bar. As a result, the upper rack and two bell cranks are held in their forward position, even though the spring-loaded bell crank is under tension. During the locking operation, the unlocking gear train is carried forward to mesh with the unlocking rack.

3-85. Unlocking the spinning and unlocking mechan-ism (figure 3-10) occurs within 0.3 second after firing the torpedo. Air-flask pressure is delivered from the starting (air) valve to the already opened impulse valve and then through nozzle ports in the front plate to rotate the spinning wheel. Rotary motion of the spinning wheel is transferred by the thrust shaft to the spinning gear meshed with the

SHAFT   SPRING LEVER   UNLOCKING BAR   SPRING ROD

WORM WHEEL AND GEAR

PINION

SPINNING GEAR

SPINNING SHAFT SLEEVE

UNLOCKING RACK

SPRING CASE HINGE

SPRING CASE

SPRING ROD

UNLOCKING TIMING ADJUSTMENT ASSEMBLY

CENTERING PIN

UPPER RACK

SPINNING GEAR FRAME

PINION

SPRING

SPRING BELL CRANK ASSEMBLY

LOCKING BAR

LOCKING LEVER

SHOCK STOP

HAND TRIP LEVER ASSEMBLY

NIPPLE

SPINNING WHEEL

VALVE

FRONT PLATE

SWIVEL BLOCK

VALVE BELL CRANK ASSEMBLY

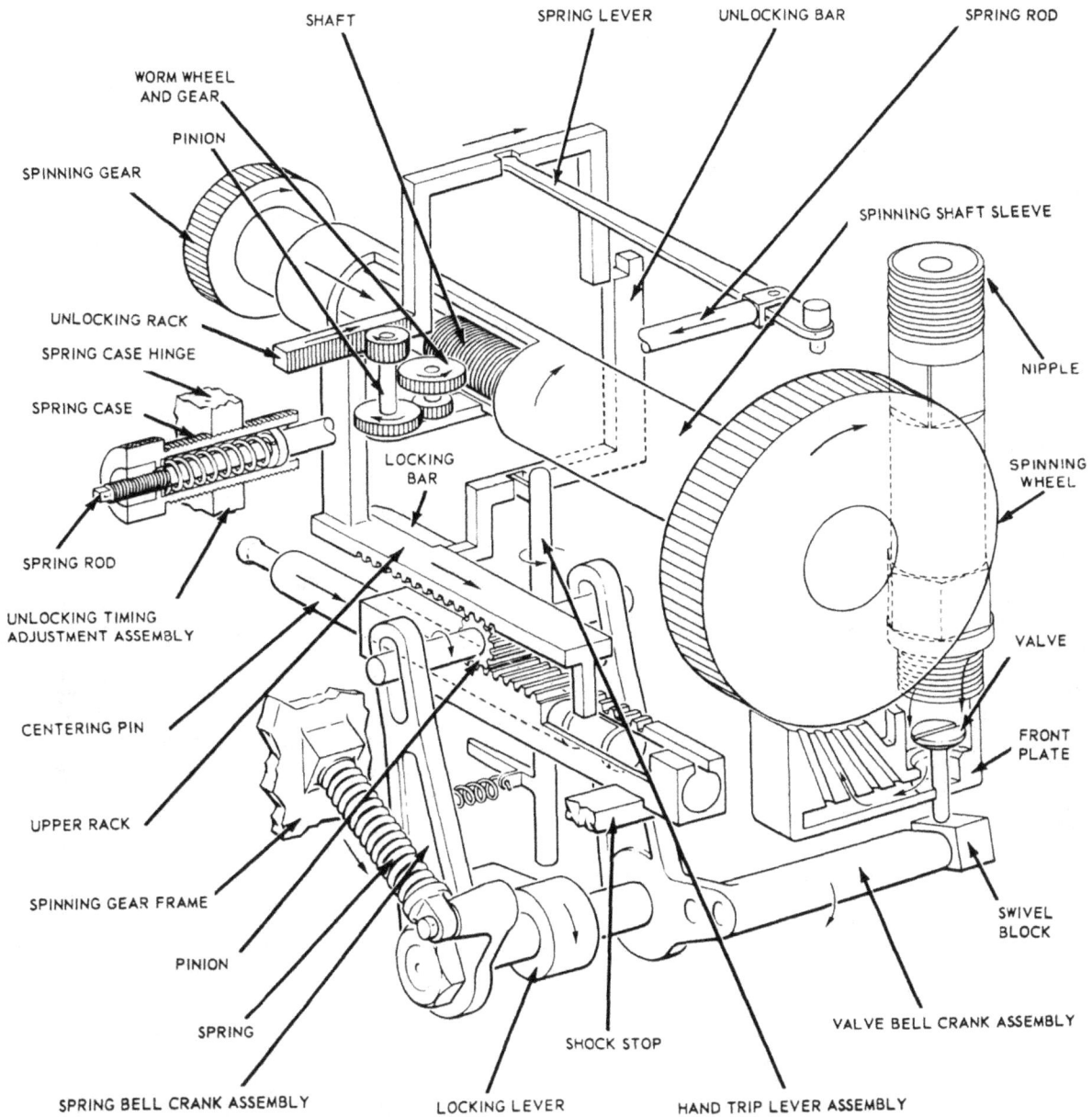

Figure 3-10. Unlocking Operation of Spinning and Unlocking Mechanism.

gyro wheel gear. Because the spinning gear and the gyro wheel gear have the same number of teeth, their speed of rotation is the same. Air-flask pressure accelerates the spinning wheel at a high-speed, and, after 20 to 28 turns of the spinning wheel, the gyro wheel is spinning at full speed. A worm cut in the forward end of the thrust shaft drives the gear train that unlocks the mechanism. As the shaft rotates, the worm and gear train drive the unlocking rack to the right. At a preset number of turns, the unlocking rack toe engages the upper end of the unlocking bar and pushes it to the right. The lower end of the unlocking bar moves from behind the locking bar and releases the upper rack. The upper rack is

then driven aft by the pinion mounted between the two bell cranks and by tension of the spring-loaded bell crank. At the same time, the spinning wheel, shaft, and spinning gear are driven aft by the upper rack. During the early portion of aft travel of the upper rack, the centering pin remains engaged with the locking end bearing of the gyro inner gimbal ring. The delay of centering pin disengagement is permitted by the recessed diameter at the aft end of the centering pin, which allows a movement of approximately 3/8-inch before the upper rack toe moves the centering pin. Thus, the spinning gear and gyro wheel gear are unmeshed before the centering pin is disengaged from the locking end bearing of the gyro inner gimbal to prevent spinning gear torque from deflecting the gyro. As the bell cranks move aft, the swivel block on the valve bell crank moves down to release the impulse valve; then the combined forces of air-flask pressure and gravity close the impulse valve and shut off air delivery to the nozzle ports in the front plate. At the completion of the unlocking operation, the gyro is free, the spinning wheel is clear of the nozzle ports, and the unlocking gear train is disengaged from the unlocking rack.

3-86. The duration-of-spin (unlocking time) adjustment is provided to insure that the locking operation remains in effect for the time required to bring the gyro spin up to 12,000 rpm. Approximately 24 turns of the spinning wheel are necessary to obtain the required unlocking time (about 0.3 second). The adjustment allows the varying length of unlocking rack travel to delay the unlocking bar movement required to disengage the locking bar that holds the upper rack in its forward position. The duration of unlocking time can be adjusted only when the spinning and unlocking mechanism is in an unlocked condition. In this condition, the unlocking gear train is disengaged from the unlocking rack. The adjustment is made by turning the knurled knob of the spring case. Because the spring case hinge is fixed, clockwise rotation of the knob draws the spring rod, spring lever, and attached unlocking rack to the left and increases the distance from the unlocking bar. Consequently, when the spinning and unlocking mechanism is cocked, more turns of the unlocking gear train are required to disengage the unlocking bar and a longer time is required for the spinning gear to become unmeshed from the gyro wheel gear. In an opposite manner, counterclockwise rotation of the knob results in fewer turns of the unlocking gear train and subsequent shorter time of spinning gear engagement with the gyro wheel gear. Altering the adjustment is prevented by spring tension between the spring rod and knob, which acts as a friction brake to prevent the spring case from turning by influence of vibration.

3-87. Pallet Mechanism. The pallet mechanism (figure 3-11) operates in various modes. When the torpedo is running on a preset course, the cam in the cam plate of the gyro will be in axis alinement with the two cam pawls. The oscillating pallet slide carries the pallet, pallet shaft, and cam pawls in a continuous fore and aft motion. As this group of components moves forward, the cam pawls momentarily contact the gyro cam plate. The space between the toes of the cam pawls permits the pawls to straddle the cam of the gyro cam plate with the lower or right-hand pawl fitting into the lower cam recess. Any deviation between the cam pawls and the pallet shaft is corrected by the momentary contact of one of the pawls with the gyro cam surface. As a result, the pallet on the upper end of the pallet shaft moves aft and is held in this position by the leaf spring on the pallet shaft. The oscillating drive assembly immediately carries the cam pawls aft and out of contact with the gyro cam plate. In this condition, no course order is transmitted to the vertical steering engine.

3-88. When the torpedo course varies to either side of the gyro angle axis, the pallet shaft is carried away from exact alinement with the gyro cam plate. On the next forward movement of the pallet slide, one cam pawl contacts the gyro cam surface and the pallet shaft is rotated so that the upper pallet moves to the left or right in relation to any change in torpedo course. If the deviation in torpedo course is great, the point of cam pawl contact will be beyond the gyro cam surface. In case of a great amount of deviation, the upper ridge of the gyro cam plate strikes the appropriate lower cam pawl to initiate a correction in course error. On the next movement of the oscillating drive, the pallet slide carries the pallet assembly in an aft direction and the pallet shaft, pawls, and upper pallet assume a new position.

3-89. The pallet pawls are affected by the oscillating fore and aft movement of the upper pallet. The pallet pawls pivot on the posts of the pallet slide cover and, because of their interconnecting linkage, the movement of one pallet pawl is in an equal amount and opposite direction to the movement of the other pallet pawl. The left-hand pallet pawl has an extra post connected to the bell crank of the pallet slide cover by the adjustable eye connection. The fore and aft movement of the left-hand pawl cause an up-and-down movement of the bell crank. The bell crank has two pins riding in the lower external groove of the connection spool to cause up-and-down movement of the connection spool. The upper external groove of the connection spool receives the lug on the arm of the valve rock shaft assembly to transmit the up-and-down movements to the valve of the vertical steering engine.

3-90. When the torpedo course is true, the pallet pawl positions will be in alinement and the linkage to the vertical steering engine will be in a neutral position, whereby the valve of the vertical steering engine is not affected. In this condition, the upper pallet is in an after position and the aft movement of the oscillating drive carries the upper pallet between the pawls without affecting them. When the torpedo varies from true course, the cam pawls no longer straddle the gyro cam and the next forward movement of the oscillating pallet slide will cause the upper pallet to rotate right or left. When the pallet slide moves aft, the upper pallet contacts one of the pallet pawls and swings it aft. Depending on which pallet pawl swings aft, the bell crank of the pallet slide cover will cause up-or-down movement of the connection spool and consequent operation of the valve rock shaft assembly and vertical steering engine. The vertical rudders oscillate to the left and right because the fore-and-aft movement of the pallet slide cover is continuously causing an up-and-down movement to the connection spool and valve rock shaft assembly. At the same time, the cam pawls pass the gyro cam and assume a neutral position, and, although the upper pallet moves between the pallet pawls, it will not cause them to assume a neutral position. Consequently, the last signal from the pallet mechanism through the valve rock shaft assem-

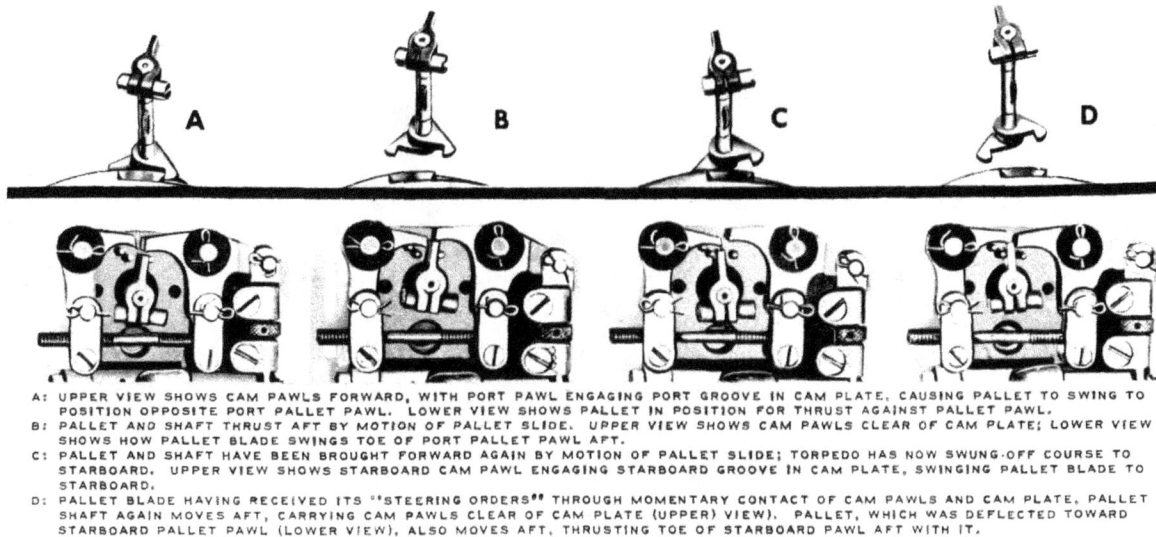

A: UPPER VIEW SHOWS CAM PAWLS FORWARD, WITH PORT PAWL ENGAGING PORT GROOVE IN CAM PLATE, CAUSING PALLET TO SWING TO POSITION OPPOSITE PORT PALLET PAWL. LOWER VIEW SHOWS PALLET IN POSITION FOR THRUST AGAINST PALLET PAWL.
B: PALLET AND SHAFT THRUST AFT BY MOTION OF PALLET SLIDE. UPPER VIEW SHOWS CAM PAWLS CLEAR OF CAM PLATE; LOWER VIEW SHOWS HOW PALLET BLADE SWINGS TOE OF PORT PALLET PAWL AFT.
C: PALLET AND SHAFT HAVE BEEN BROUGHT FORWARD AGAIN BY MOTION OF PALLET SLIDE; TORPEDO HAS NOW SWUNG-OFF COURSE TO STARBOARD. UPPER VIEW SHOWS STARBOARD CAM PAWL ENGAGING STARBOARD GROOVE IN CAM PLATE, SWINGING PALLET BLADE TO STARBOARD.
D: PALLET BLADE HAVING RECEIVED ITS "STEERING ORDERS" THROUGH MOMENTARY CONTACT OF CAM PAWLS AND CAM PLATE, PALLET SHAFT AGAIN MOVES AFT, CARRYING CAM PAWLS CLEAR OF CAM PLATE (UPPER) VIEW). PALLET, WHICH WAS DEFLECTED TOWARD STARBOARD PALLET PAWL (LOWER VIEW), ALSO MOVES AFT, THRUSTING TOE OF STARBOARD PAWL AFT WITH IT.

Figure 3-11. Pallet Mechanism Operation.

bly to the vertical steering engine remains in effect until an off-course signal in the opposite direction is initiated by the ridge on the gyro clamp plate striking the appropriate lower cam pawl to change vertical rudder position. Because of the movement of the vertical rudders, the torpedo is continually "off course," and, in trying to make a correction, the vertical rudders are regularly alternating port and starboard; this results in a slight zigzag during the run.

3-91. Vertical Steering Engine. When pallet mechanism operation causes the valve rock shaft assembly to move the pilot valve aft, working air pressure is ported into the after end of the piston chamber; the piston is forced forward to produce a starboard rudder position (figure 3-12); and residual pressure at the forward end of the piston chamber is vented through the applicable exhaust port into the afterbody. When the piston reaches its full forward position, it produces a hard-over starboard rudder position.

3-92. In the opposite steering operation, working air pressure is ported into the forward end of the piston chamber; the piston is forced aft to produce a port rudder position (figure 3-12), and residual pressure is vented through the applicable exhaust port into the afterbody. When the piston reaches its full aft position, it produces a hard-over port rudder position.

3-93. DEPTH CONTROL. The immersion mechanism, in conjunction with other components, maintains the running depth. The depth-setter assembly sets the immersion mechanism depth spring to a tension equivalent to sea-water depth pressure at the desired running depth. At the desired running depth, actual sea-water depth pressure is offset by the depth spring tension to equal atmospheric pressure at sea level (the reference), and the result is compared to sea-level air pressure in the air chamber and diaphragm assembly. Any imbalance is transmitted to

the horizontal steering engine, which corrects for deviations sensed by the diaphragm. Broaching is prevented by the climb-angle limiter device.

3-94. Depth-Setter Assembly. Unlike the gyro servosystem, the depth servosystem (figure 3-4) operates at only a high, single speed. Although low-speed synchro control transformer SCT-L is provided in the relay transmitter of the fire control system and single low-speed relay RY3 is provided in the torpedo amplifier unit, these components have no application in the setting of the depth order.

3-95. When operating with the fire control system, the stator windings of synchro transmitter SG3 are interconnected with the stator windings of high speed (300 feet per turn) synchro-control transformer SCT-H in the relay transmitter of the fire control system. A reference voltage to energize the rotor of synchro transmitter SG3 is obtained from one phase of the firing craft 115-v 400-Hz three-phase power source.

3-96. Servomotor M2, a two-phase induction motor, drives the gear train. Its main field is energized by a constant amplitude voltage from one phase of the firing craft 115-v 400-Hz three-phase power source. Its control field is energized by the variable amplitude of the 400-Hz voltage, which is the amplified error voltage from relay transmitter synchro control transformer SCT-H. The two field voltages must have a phase difference of 90°. Resistor R2 and capacitor C2 of the R-C network are shunted across the control field of servomotor M2 to provide additional phase shift to the control field voltage and to insure a quadrature with the main field voltage.

3-97. Upon being driven by servomotor M2, the gear train drives the gear on the adjusting screw for the depth spring. When synchro control transformer SCT-H receives a depth-set order from the fire control system, it causes servomotor M2 to drive the

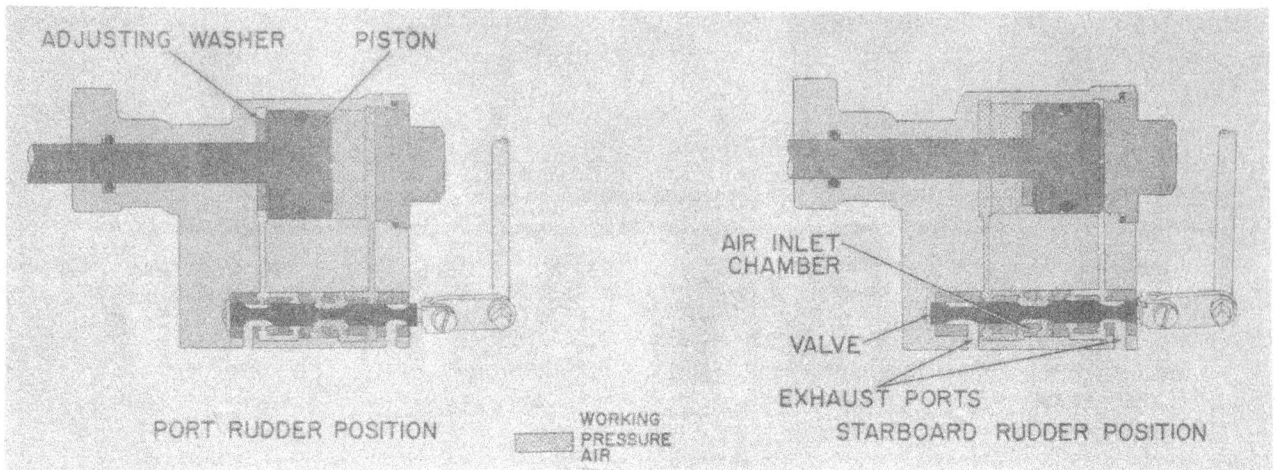

Figure 3-12. Vertical Steering Engine Operation.

gear train in the direction required to correctly set the adjusting screw for the depth spring. Servomotor M2 also drives synchro transmitter SG3 in correspondence with synchro control transformer SCT-H to cancel the error order signal.

3-98. The depth setter does not have a solenoid brake. However, if a depth-set order in excess of 50 feet is attempted, the depth-set servo order system is disabled at the 50-foot setting by the depth-limit cam, limit switch, and limit relay in the fire control system. Thus, an attempt to set an excess of 50 feet will result in the depth setter producing a 50-foot setting of the depth spring until a shallower setting is made.

3-99. Immersion Mechanism. A combination of sea-water depth pressure and torpedo noseup or nosedown attitude controls operation of the immersion mechanism (figure 3-13) to maintain the torpedo at a preset running depth. When the torpedo leaves the tube, sea water enters the water chamber and depth pressure is applied to the diaphragm and depth spring. The pendulum swings, in combination with the amount of sea pressure applied on the diaphragm and depth spring and with the torpedo noseup and nosedown attitudes. When the torpedo approaches the set depth, the pendulum, diaphragm, and depth spring cause the horizontal steering engine to produce a gradual neutral position of the horizontal rudders. Soon after the torpedo run is started, the neutral position of the horizontal rudders is acquired and maintained by combined action of the pendulum swing and sea water pressure on the diaphragm and depth spring, as they affect the horizontal steering engine. If the torpedo is running nosedown or too deep, the pendulum swings forward, the diaphragm and depth spring move down, and the linkages operate in accordance with the combined forces to cause the horizontal steering engine and rudder rod connections to produce an up-rudder position. If the torpedo is running noseup or too shallow, the pendulum swings aft, the diaphragm and depth spring move up, and the linkages operate in accordance with the combined

forces to cause the horizontal steering engine and rudder rod connections to produce a down-rudder position.

3-100. Before a torpedo run, the internal air-chamber pressure is equalized with atmospheric pressure at sea level and then is sealed by an O-ring and plug installed in the bottom of the air chamber to keep the atmospheric pressure constant throughout the run. Pressure balance between the atmospheric pressure in the air chamber and sea-water depth pressure in the hollow cylindrical bottom of the immersion gear casing is affected by variations in atmospheric conditions and temperature of air and water. Therefore, actual depth of run will differ from set depth, unless both factors are accounted for in the amount of diaphragm movement. The opposing differential between air and water pressure on the diaphragm, plus the tension of the depth spring, determines the actual depth of the run. Corrections for the opposing pressures (see volume 2) must be applied using an indicated depth setting somewhat different than the desired depth of run. The combined diaphragm and depth spring movement through the connecting linkage moves the pilot valve of the horizontal steering engine in a greater amount with a ratio of 18 to 1. The diaphragm can be set to attain a position required for a depth spring tension setting equivalent to a 50-foot running-depth setting.

3-101. Horizontal Steering Engine. The horizontal steering engine (figure 3-14) receives steering orders from pendulum operation as influenced by the amount of deviation between the actual running and preset depth. It produces the force necessary to position the horizontal rudders and controls the degree, as well as direction of rudder throw. It is a followup type of engine controlled by a sensitive internal valve. It operates on the working air pressure supply received from the afterbody test connection.

3-102. The working air pressure passes through the air strainer, enters the cavity between the piston O-ring bearing surfaces, and passes through the

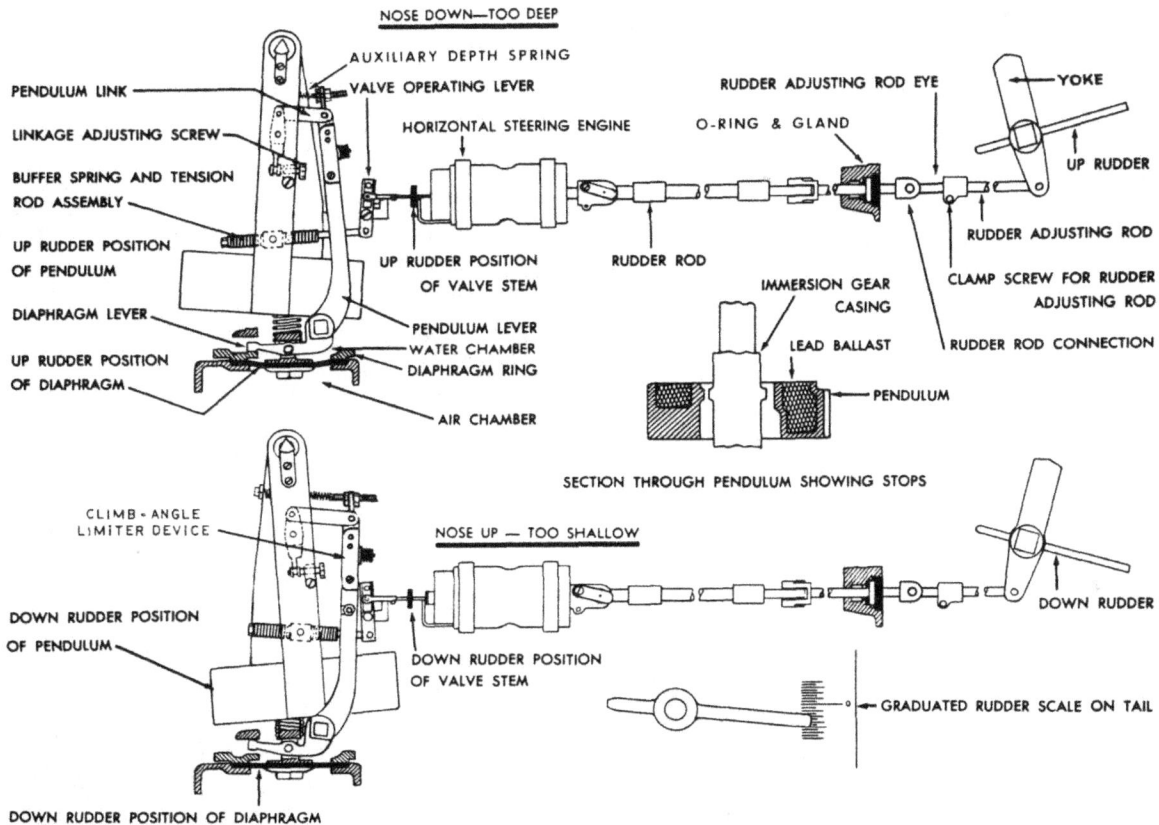

Figure 3-13. Immersion Mechanism Operation.

air inlet hole in the piston body to the cavity between the pilot valve and the piston. When a neutral depth steering order is given, the working air pressure is banked in the cavity because the land bearing surfaces of the pilot valve blank off both piston one-way ports (top view of figure 3-14). When a down-rudder order for depth steering is given, the pilot valve is moved forward; the forward land bearing surface of the pilot valve uncovers the forward one-way port in the piston; and working air pressure enters the aft end of the cylinder body. Simultaneously, the after land bearing surface of the pilot valve uncovers the after one-way port in the piston; the residual pressure in the forward end of the cylinder body is vented through the pilot valve and after-piston shaft into the afterbody; and the piston is forced forward (center view of figure 3-14). When the piston movement overtakes the pilot valve movement, the land bearing surfaces of the pilot valve close both one-way ports of the piston to stop working air pressure delivery to the after end of the cylinder body and to stop residual air in the forward end of the cylinder body from venting; thus, no further movement of the piston results. When an up-rudder order for depth steering is given, the pilot valve is moved aft and the operation of the horizontal steering engine is reversed (bottom view of figure 3-14).

3-103. Antibroach. The climb-angle limiter device provides the pendulum lever with a scissor type of

action under a spring tension, primarily to limit and maintain the transmitted depth-error torque to any preset maximum and to avoid a steep climb angle and broach at the start of the run. Under an ideal situation, such as when the torpedo is launched at a depth slightly deeper than its set-to-run depth, the depth error difference between the depth diaphragm and the depth spring is transmitted from the diaphragm through the pendulum lever to swing the pendulum forward. Subsequent operations of the related linkages, horizontal steering engine, and rudder rods produce an up-elevator position and a resulting up-attitude and climb-pitch angle of the torpedo. Up-elevator position is held until the depth control elements of the depth spring tension versus hydrostatic pressure on the depth diaphragm balance to cause a neutral rudder position and a resulting set depth run.

3-104. Under an adverse situation, as when the torpedo is launched deep with a shallow-running depth setting and without the use of the climb-angle limiter device, the high force of the hydrostatic pressure will hold the elevators at a hard-up position and cause the torpedo to climb increasingly steeper to the extent that, by the time a neutral elevator position can be attained, the torpedo has climbed above its set depth and broaches. In this same situation, but with the climb-angle limiter device in use, the torpedo will attain the set running depth without the danger of a

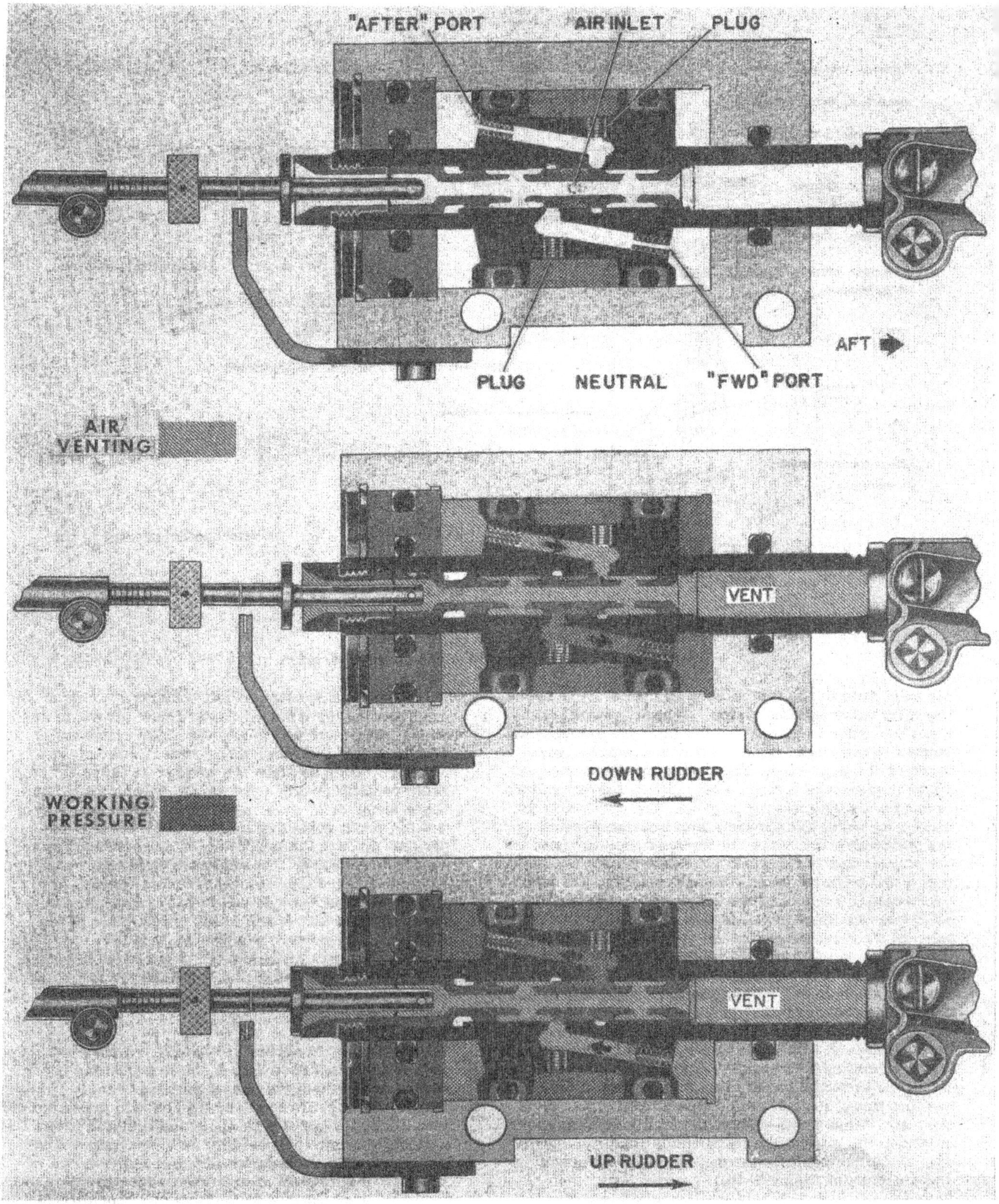

Figure 3-14. Horizontal Steering Engine Operation.

broach. For example, when the torpedo is launched at a 180-foot depth without use of the climb-angle limiter device, it will acquire a high climb-pitch angle of 19° to 21° and a resulting overshoot of the set depth. Under similar conditions, but with the use of the climb-angle limiter device, the torpedo will acquire a climb-pitch angle of only 10° to 10.5° without an overshoot.

3-105 thru 3-110. Deleted by CHANGE 2

3-111. ENERGY CONTROL. The energy section is covered in paragraph 3-42 along a flow-path basis; in the following paragraphs, the various switches and valves used in the energy section are described in detail. Figure 3-15 simplifies the pneumatic and hydraulic system to emphasize these switches and valves.

3-112. Stop and Charging Valve. The stop and charging valve (figure 3-16) is a dual valve, manually operated to charge the air flask (both valves open) and confine air pressure within the flask (both valves closed). A conical steel tip on the charging line holds the charging valve open when the flask is being pressurized, and spring action reseats the valve when the charging line is removed. Before removing

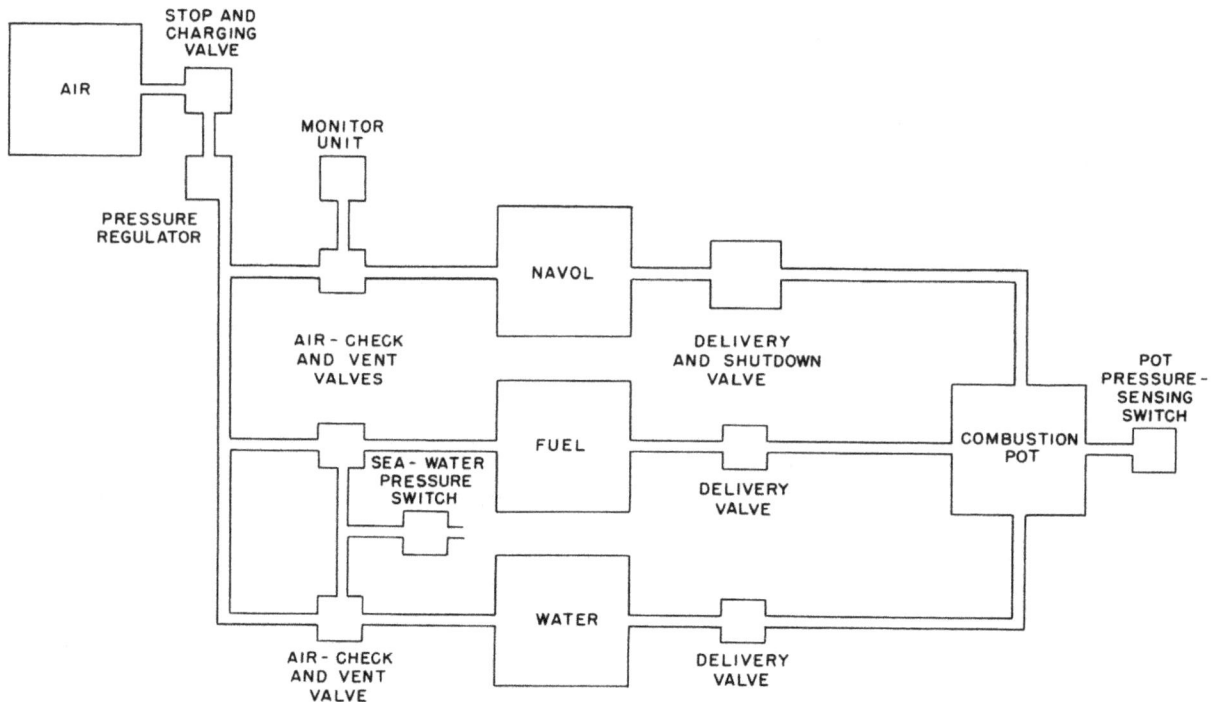

Figure 3-15. Pneumatic and Hydraulic System, Simplified Flow Diagram.

the charging line, the stop valve is closed to confine pressure. Reopening the stop valve allows air flask pressure to bank at the starting valve, through nipple B, ready for use when the torpedo is fired. With the stop valve closed and the lines bled, the afterbody may be separated from the charged air flask. If the afterbody is not assembled to the air flask, nipple B is blanked off while the flask is being charged. A warning sign is installed to indicate that the air flask is pressurized.

3-113. Air-Check and Vent Valve. When working pressure air is delivered to each air-check and vent valve, it is admitted to the inlet side of the air-check valve portion of the valve body, passes through the air-check valve seat and is piped behind the vent valve, which is seated. Then, the air passes through an outlet chamber to be delivered to the Navol tank, water compartment, or fuel tank. At the same time, a portion of the air is ported through a passage in the valve body to a chamber at one end of the vent valve piston where the pressure differential is great enough to overcome the spring tension on the piston and vent valve. Because the area at the piston end is greater than the area surrounding the vent valve, the working pressure air in the outlet chamber is insufficient to open the vent valve and all admitted air is directed to the void space over the applicable liquid. When the torpedo is in a fully ready condition and after a torpedo run, working pressure does not exist, and there is little or no pressure in the

chamber at end of the vent valve piston. As a result, excessive pressure in the Navol tank, water compartment, and fuel tank enters each applicable valve in a reverse direction of that for delivery of working pressure air. Venting pressure enters the valve body chamber between the air-check valve and the vent valve. The fuel and water air-check valves remain closed, because they are under a spring tension applied in the same direction as the venting pressure. Their vent valves open when the venting pressure becomes great enough to overcome spring tension. The Navol air-check valve differs from the fuel and water air-check valves by being normally open and closing only upon application of working air pressure.

3-114. Sea-Water Pressure Switch. The sea-water pressure switch, mounted on the fuel and water overboard vent opening, prevents premature torpedo firing. Normally open contacts of the switch are connected by cable to the power-pack firing circuit. Before the torpedo can be fired, a sea-water pressure of 4 ±1 psi equivalent to a sea depth of 10 feet must exist at the inlet of the fuel and water overboard vent opening to close the switch contacts. With the switch contacts closed, the firing circuit in the power pack can receive the firing impulse voltage.

3-115. Navol Delivery and Shutdown Valve. Before the torpedo is fired, the normally open contacts of the sea-water pressure switch must be closed to

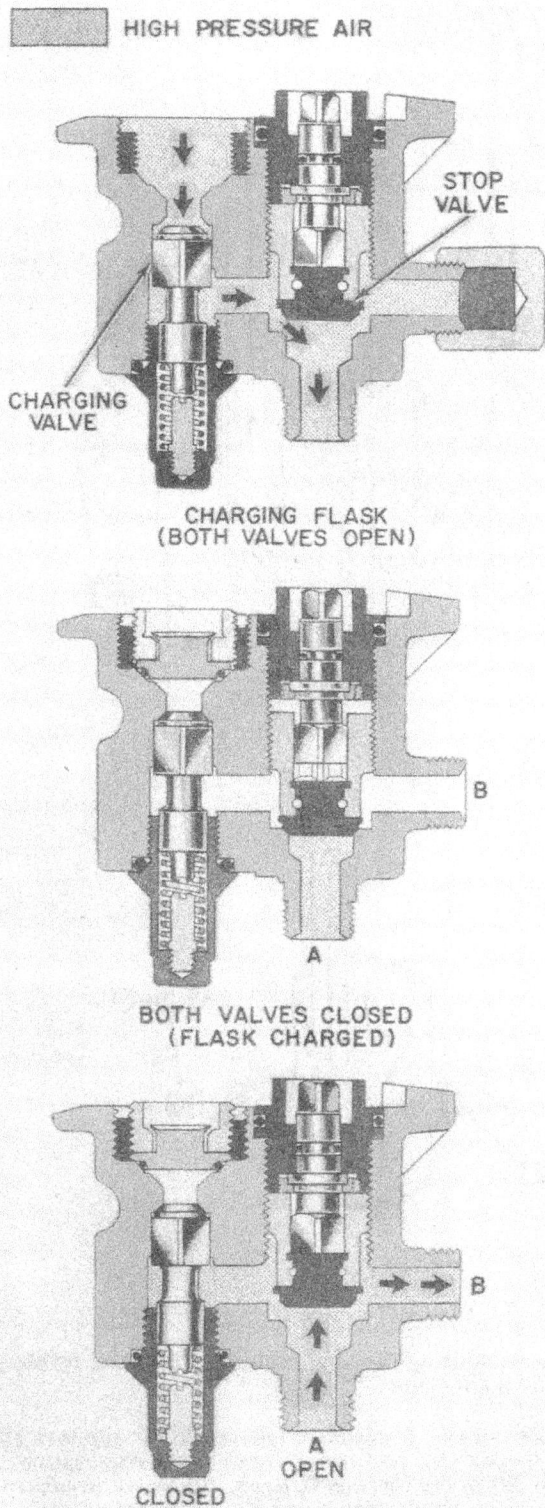

HIGH PRESSURE AIR

STOP VALVE

CHARGING VALVE

CHARGING FLASK
(BOTH VALVES OPEN)

BOTH VALVES CLOSED
(FLASK CHARGED)

OPEN

CLOSED

Figure 3-16. Operation of the Stop and
Charging Valve.

allow the power pack to deliver a signal voltage to the Navol delivery and shutdown valve. When the firing impulse voltage is applied, the signal voltage from the power pack fires the applicable squibs to cause the opening of the valve and delivery of Navol to the combustion system. Two squibs, used for the delivery operation, are wired in parallel and are fired by a common current to insure that the valve opens. For Navol shutdown operation, two other squibs are fired by the power pack under the control of the governor overspeed switch, anti-circling-run device, or pot pressure switch. One squib is wired in series with the contacts of the overspeed switch circuit, which in turn is wired in parallel with the anti-circling-run circuit. The other squib is wired in series with the contacts of the pot pressure sensing switch. When any one of the shutdown functions is initiated the required squib is fired, causing the valve to close, and Navol delivery ceases.

3-116. Pot Pressure-Sensing Switch. When the pressure in the combustion system reaches 400 $^{+30}_{-0}$ psi at the start of the torpedo run, the pot pressure-sensing switch is actuated and opens the applicable power pack circuit used for low pot-pressure shutdown. When the pot pressure drops below 400 psi, such as during a malfunction of the combustion system and at end of the torpedo run, a power pack circuit used for low pot-pressure shutdown is completed by the closed contacts of the switch, causing squib activation of the shutdown portion of the Navol delivery and shutdown valve.

3-117. Navol Monitoring Unit. When the monitoring unit (figure 3-17) is operating, gases resulting from decomposing Navol enter the inlet fitting, pass through the U-shaped tube, through a bubble-metering device in the common chamber, into the body compartment, and out the check valve. As the gas passes through the bubble-metering device, it is in the form of bubbles. When each bubble passes the probe, the probe is momentarily wiped dry to produce a change in resistance used in obtaining a signal sent by way of the monitoring unit cable and Navol surveillance and power cable to the indicator panel. Sodium nitrate is added to the distilled water so that a resistance of 3,500 to 60,000 ohms exists in the signal path before the probe is wiped by a bubble. When the probe is wiped by a bubble, the signal path is interrupted and recorded on the indicator panel. The bubble rate is used to determine decomposition rate of Navol during surveillance.

3-118. DETAILED FUNCTIONAL DESCRIPTION OF PROPULSION SYSTEM

3-119. COMBUSTION SYSTEM. The combustion system generates power to drive the turbine and gear train assembly (main engine) of the torpedo. It generates power by converting the elements contained in the Navol, alcohol, and water into hot gases and steam, which are directed at high velocity to the turbines of the main engine.

3-120. The decomposition chamber decomposes the Navol into hot oxygen and steam by catalytic action and applies a water spray to act as a coolant for the

WATER ACCESS HOLE (2)

A

A

B

B-B

GAS INLET

PROBE

WATER ACCESS HOLE (2)

GAS INLET

PROBE

A-A

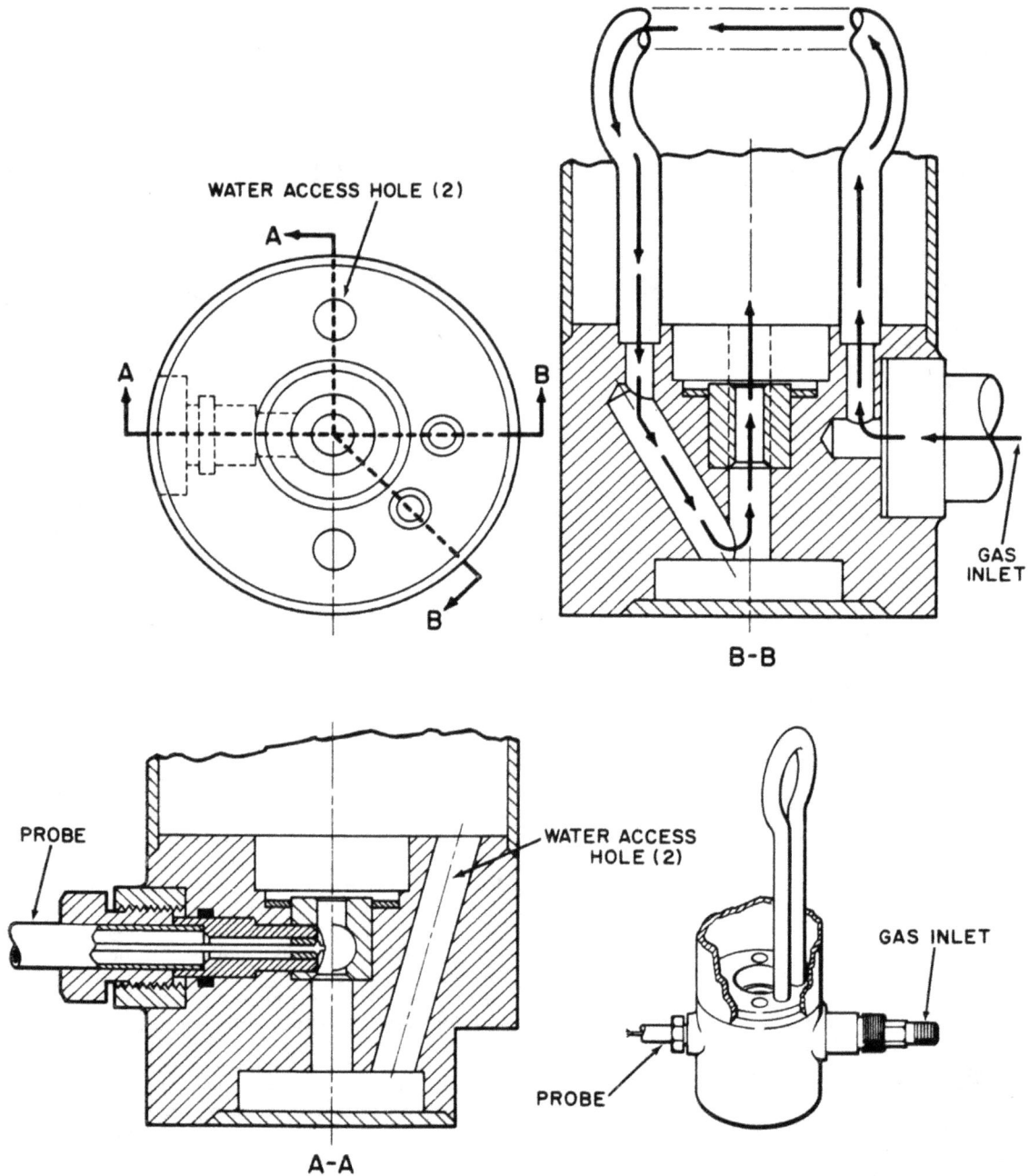

Figure 3-17. Navol Monitoring Unit Operation.

decomposed Navol and eventually to become additional steam. The combustion pot mixes hot oxygen, steam, and water spray from the decomposition chamber, combines the mixture with alcohol, and the final mixture is ignited, thereby generating a high pressure made up of hot gases and steam.

3-121. The nozzle, pipe, and combustion pot bottom receives the high-pressure hot gases and steam produced in the combustion pot, which it delivers at

high velocity against the buckets of the first turbine of the main engine.

3-122. Water pressure triggers the two igniters at the instant that fuel is delivered to the combustion pot. When the torpedo is fired, the water pressure enters the inlet opening and is applied on the diaphragm with sufficient force to activate the spring housing assembly, which drives the two firing pins through the shearing plate, setting off the two primer caps and three charges of ignition load. When the ignition loads are set off, they burn with enough heat to destroy the perforated disk and metallic seal at the igniter outlet opening and long enough to insure ignition of the fuel (alcohol) in the combustion pot.

3-123. TURBINE AND GEAR TRAIN. The first turbine wheel and spindle assembly is driven in a counterclockwise rotation (looking forward) by the angular flow of high-velocity hot gases and steam ejected from the nozzle of the combustion system. The driving power of the first turbine wheel is transmitted from its pinion gear through the countershaft and gearing assembly and after idler gear assembly to drive the forward propeller shaft and gear assembly and also through the countershaft and gearing assembly alone to drive the after propeller shaft and gear assembly.

3-124. The second turbine and spindle assembly is driven in a clockwise rotation (looking forward) by the angular flow of high-velocity hot gases and steam ejected from the buckets of the first turbine wheel. The driving power of the second turbine wheel is transmitted from its pinion gear through the forward idler gear assembly, countershaft and gearing assembly, and after idler gear assembly to drive the forward propeller shaft; and also through the forward idler gear assembly and countershaft and gearing assembly to drive the after propeller shaft.

3-125. DETAILED FUNCTIONAL DESCRIPTION OF WARHEAD

3-126. Detonation of the main explosive charge is initiated by the exploder as contact or influence firing. In contact firing, the exploder is triggered by the impact of the torpedo against the hull of the vessel. In influence firing, the exploder is triggered by the signal voltage generated in the search coil when the torpedo passes closely beneath the hull of the vessel. (For safety, the influence-firing mode is disabled above 5.2 feet and below 92 feet.) After the exploder is activated, a small explosive charge of either powder train in the arming device is detonated by the exploder firing circuit; the powder trains and explosive (tetryl) charge in the arming device then set off the explosive (tetryl) pellets in the booster, which detonates the main explosive charge.

3-127. COMBINED INFLUENCE AND CONTACT FIRING. Exploder Mk 9 Mod 7 is armed approximately 9 seconds after torpedo launch, or at 250 yards from the launching tube. For exploder-influence firing, the search coil (figure 3-18) sends a voltage signal to the amplifier unit of the exploder when the torpedo nears the target. Then, when the torpedo is close to and beneath the target, the voltage signal increases in magnitude and is amplified sufficiently to cause the firing of the 1/3-second delay primer in the arming device. For exploder-contact firing, the voltage generated within the exploder itself causes the instantaneous primer of the arming device to fire when the inertia switch is tripped upon torpedo impact with the target. After either one of the primers is fired, the sequential order of smaller explosions of the powder trains and tetryl pellets in the arming device and the tetryl pellets in the booster detonate the main explosive charge in the warhead.

3-128. When set for combined influence-contact firing, the exploder is most likely to detonate the main explosive charge, destroying the enemy vessel. When the torpedo is fired at large ships and a near-miss or no impact occurs, the main explosive charge will be detonated and considerable damage can be inflicted to the ship, even though the torpedo passes as much as 15 feet below the keel of the ship. Without the use of the influence firing, compensation would have to be made in the running-depth setting to allow for erratic torpedo depth performance or incorrect estimate of target draft. Consequently, combined contact and influence firing allows deeper settings of running depth without concern about running deeper than target draft.

3-129. For example, when an exploder is set for contact-only firing, it would not be feasible to set a torpedo running depth of more than 10 feet for a target with an estimated draft of 10 feet. An error of 2 or 3 feet may occur in both the estimated target draft and actual torpedo running depth. Therefore, by using combined influence-contact firing, a torpedo running depth of 10 feet can be set for an estimated target draft of 8 feet because, if the actual target draft were 8 feet and the torpedo actually ran 2 feet below the set depth, the torpedo would pass within 4 feet of the keel of the target. At this distance below the keel, good results from influence firing are certain; also, the torpedo performs better at a running depth of 10 feet, because surface wave interference is reduced, the torpedo is less likely to be sighted, and greater damage is inflicted on the vessel upon detonation of the main explosive charge. In most cases, influence firing makes it possible to complete successful attacks on shallow-draft enemy vessels, whereby contact-only firing would not be feasible. In conclusion, it is recommended that the exploder be set for combined influence-contact firing and that the torpedo running depth be set at the estimated depth of the target keel or 10 feet, whichever is greater, to provide a low-contact or proximity hit.

3-130. INFLUENCE FIRING. Influence firing by the exploder depends on the torpedo entering the gradient magnetic field surrounding the enemy vessel and the search coil detecting the gradient magnetic field. The search coil acts as a gradiometer with two coils approximately 1 foot apart and connected in series opposition. When the running torpedo passes through the earth's magnetic field equal and opposite voltages are induced in the two coils of the magnetically balanced search coil, and no net voltage output results. However, when the torpedo approaches the steel hull of the vessel, the situation changes, because one of the two coils will be slightly closer to the steel hull and a slightly different voltage will be induced in the closer coil. This different voltage is quite small, but it becomes large enough after proper amplification to cause exploder detonation.

3-131. The voltage signal is amplified and impressed on the grids of two thyratrons in the exploder am-

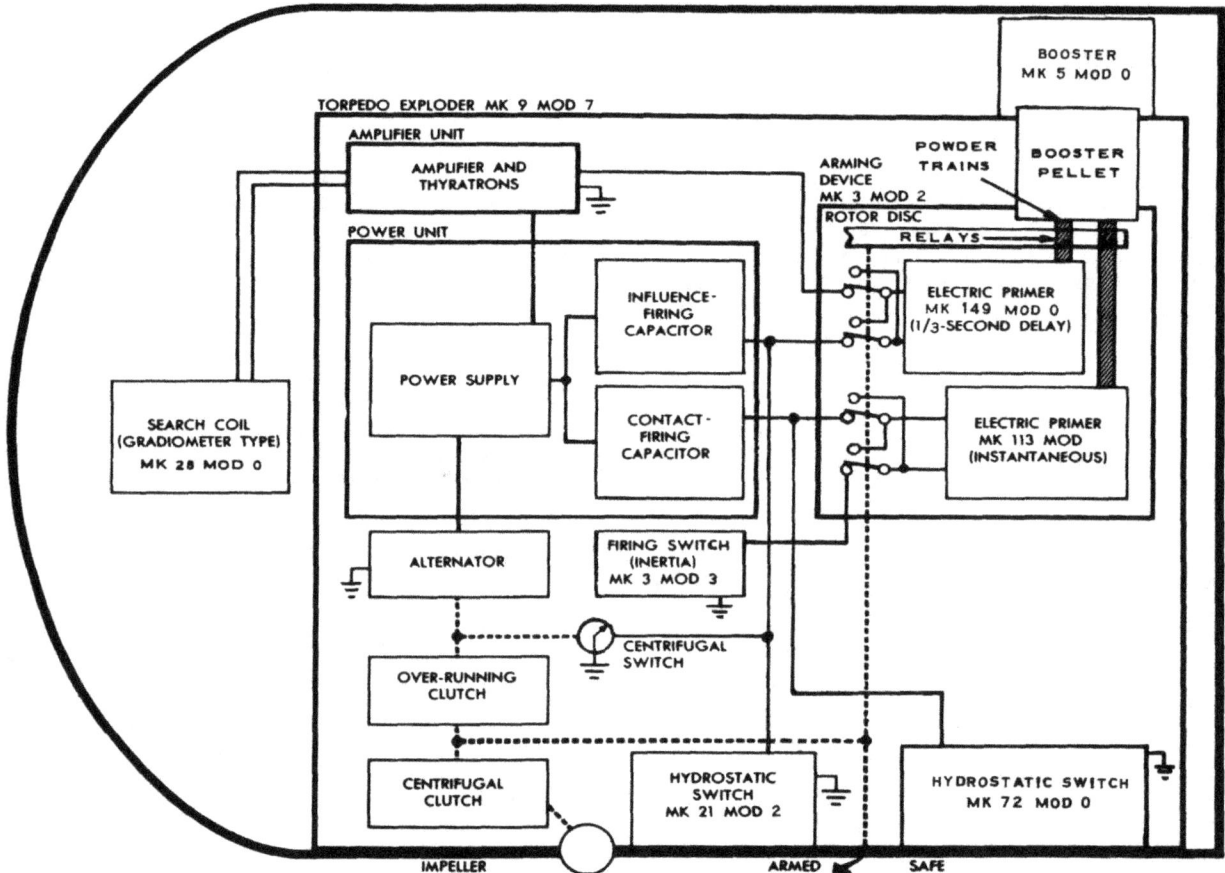

Figure 3-18. Exploder and Search Coil, Block Diagram.

plifier unit. Regardless of the polarity of the voltage signal from the search coil, one of the thyratrons will fire to discharge the influence-firing capacitor in the exploder power unit. The thyratron firing is usually triggered when the voltage induced in the search coil reaches an amount normally produced as the torpedo approaches the strongest portion of the gradient magnetic field in the vicinity of the near side of the enemy vessel hull. When the influence-firing capacitor is discharged, it sets off the 1/3-second delay primer in the exploder arming device, powder trains in the arming device, and tetryl pellets in both the arming device and booster. Thus, the detonation of the main explosive charge in the warhead occurs when the torpedo is approximately midway under the enemy vessel because the torpedo continues to travel during the elapsed time from the setting off of the 1/3-second delay primer until the detonation of the booster tetryl pellets and main explosive.

3-132. CONTACT ONLY. The use of contact-only firing by the exploder is not recommended unless a torpedo must be fired in shallow water or other tactical situations require it. Authority for setting the exploder for contact-only firing depends on the decision made by the tactical commander after he receives pertinent information from the torpedo data computer.

3-133. When contact-only firing is desired, the influence-firing circuit must be disabled by replacing the cross-slotted screw in the exploder hydrostatic switch with an influence-deactivating, single-slotted screw before or after the exploder is installed in the warhead. With the exploder set for contact-only firing, the detonation of the main explosive charge is initiated by exploder Inertia Firing Switch Mk 3 Mod 3 when the torpedo strikes the hull of the vessel. Upon impact, a spring-supported spherical weight of the inertia switch is displaced and strikes the surrounding domed surface of the inertia switch to activate the contact-firing circuit of the exploder. When the contact-firing circuit is activated, the contact-firing capacitor in the exploder power unit is discharged to set off the instantaneous primer in the exploder arming device. The remaining arming device and booster detonate the main explosive charge in the same manner as for influence firing. Complete operating time is so short that detonation occurs almost instantaneously on torpedo impact and before any appreciable premature damage can be caused to the warhead and exploder.

3-134. DETAILED FUNCTIONAL DESCRIPTION OF EXERCISE HEAD

3-135. Throughout a torpedo exercise run, the lead ballast and water ballast contribute to the proper trim

d stability of the torpedo, and the depth and roll
corder (if used) produces a permanent record of
rpedo performance. At the end of an exercise run,
e air-release mechanism pressurizes the exercise
ad to expel the water ballast through the two dis-
arge valves into the sea, thereby giving the torpedo
sitive buoyancy. The vent valve (top rear) prevents
inor accidental leakage of air pressure that might
iter the exercise head from prematurely opening
e discharge valves. If the water ballast fails to be
own or if other conditions cause the torpedo to sink
r otherwise be lost at the end of the run, the signal
om the torpedo-locating device provides for locating
e torpedo.

-136. VENT VALVE. When the ambient sea water
ressure is greater than the internal exercise head
ressure, the O-ring prevents entry of sea water
nto the exercise head. When the internal exercise
ead pressure is in excess of 3 psi because of
ccidental air leakage, the pressure is relieved by
he O-ring so that the discharge valves are not
pened prematurely.

-137. DISCHARGE VALVES. When the internal
exercise head pressure becomes greater than the
mbient sea water pressure, the discharge valves
open because a sufficient force is exerted on the
nner flanged face of each valve to overcome the com-
ined forces of each valve spring and sea water pressure against the outer flanged surface of each valve.
The water ballast is completely expelled in approximately 15 seconds. After the water ballast is expelled, it is replaced by air being expelled until
internal exercise head pressure drops to the 4 to 7
psi pressure at which the discharge valves close.
When each discharge valve closes, a rubber gasket
secured to the inner flanged face of its valve seats
on the outer flanged face of the valve body to prevent
sea water from entering the exercise head.

3-138. AIR-RELEASE MECHANISM. When air at
full air-flask pressure (2800 ±50 psi) is admitted
through the inlet nipple upon removal of the stop plug
from the blow valve, the air-release mechanism
(figure 3-19) self-cocks when the spring tension on
its valve is overcome, and the valve seats tightly.
The air-release mechanism remains cocked for as
long as air-flask pressure stays above the spring-
tension setting for its uncocking operation. When the
air-flask pressure decreases to 1450 ±25 psi, as at
the end of a torpedo exercise run, the air-release
mechanism uncocks when its valve becomes unseated
by the greater force of its spring-tension setting.
Upon unseating of the valve, air-flask pressure
(1450 ±25 psi and below) enters the inlet nipple,
passes through the filter, passes between the valve
and valve seat, and out through the outlet ports of
the main housing into the exercise head. When the
internal exercise head pressure becomes greater
than the combined sea-water pressure and valve
spring tension on the discharge valve, the water
ballast is expelled from the exercise head through
the discharge valves into the sea.

3-139. OPERATIONS

3-140. FIRE CONTROL CUTOFF. When the torpedo
is launched from the tube, all signals from the submarine fire control system are cut off by the cable
cutter attached to the A-cable plug installed in the
afterbody receptacle. Torpedo forward motion
removes A-cable slack, thus providing tension in an
aft direction on the cable-cutter arm. When the tension is sufficient, the shear pin of the cable cutter is
sheared and the rotary motion of the cable-cutter
blade shears the cable close to the plug connector and
flush with the afterbody receptacle.

3-141. NORMAL SHUTDOWN. Within 10 seconds
after the start of the torpedo run, the combustion pot
pressure reaches 400 psi and causes pot pressure
sensing switch S3 (figure 3-7) to open and remain open
throughout the normal run. At the end of the normal
run, the combustion pot pressure decreases below
400 psi; pot pressure sensing switch S3 closes, and
capacitor C4 discharges to fire squib "B" of Navol
shutdown valve V5. When squib "B" is fired, the
Navol shutdown valve closes to stop the flow of Navol
to the decomposition chamber. Eventually, the Navol
tank pressure drops sufficiently to open the vent
portion of the Navol air check and vent valve and
thereby allow the remaining air pressure and residual
Navol in the Navol tank to be expelled overboard
through the check valve of the monitor unit until only
1 to 5 psi pressure remains in the Navol tank.

3-142. At the conclusion of a normal exercise run,
the water- and fuel-delivery valves remain open and
the vent portion of the water and fuel air-check and
vent valves open to allow all remaining air pressure
and residual water and fuel to be expelled overboard
from their compartments through the exhaust system
and vent manifold. Also, when air-flask pressure
decreases to 1450 psi, the air-release mechanism
in the exercise head opens, air-flask pressure flows
into the head, and water ballast is expelled overboard through the expulsion valves.

3-143. When the torpedo run is expended, the torpedo assumes a tail-down position because the torpedo is heavier at its tail end. As a result, the
working air pressure inlets and the liquid outlets of
the liquid compartments are in a position to allow
the liquids to be expelled ahead of the air. Venting
of the liquids provides additional buoyancy.

3-144. EXCESSIVE-SPEED SHUTDOWN. An excessive-speed shutdown occurs when the torpedo
makes a slow getaway from the launching tube, during
high torpedo broaching in a run, when the torpedo
runs ashore, and in event of an accidental start when
loading the torpedo into the tube. In most of these
cases, the torpedo propellers would be out of water,
thereby allowing the torpedo main engine to develop
an overspeed because of insufficient load on the
propellers. In all cases, the main engine overspeed
causes the governor to operate as required to initiate
a shutdown.

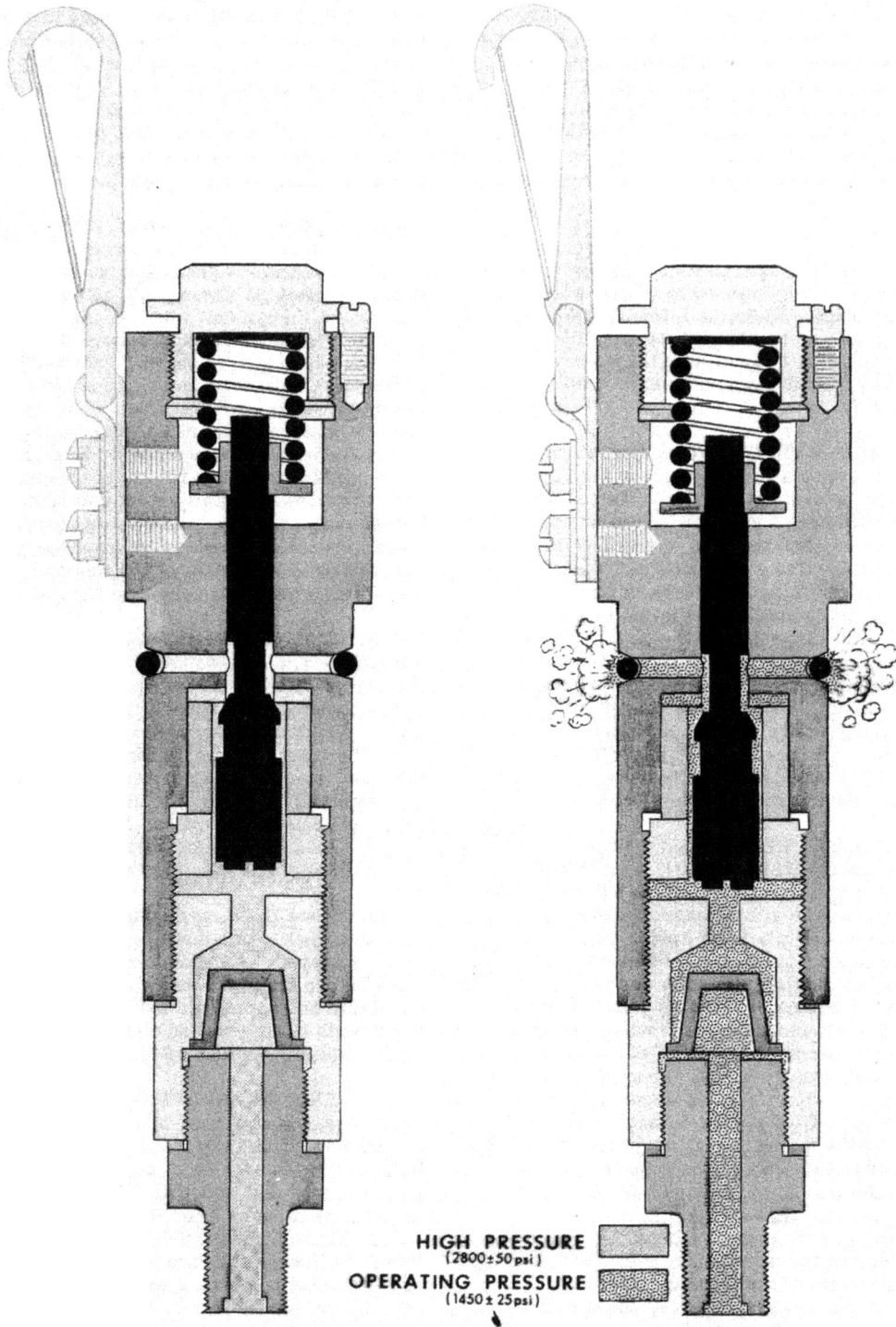

HIGH PRESSURE
(2800±50 psi)

OPERATING PRESSURE
(1450± 25 psi)

Figure 3-19. Air-Release Mechanism Operation.

3-145. When the main engine overspeeds, the governor spins more rapidly than normal. The increased rate of governor spin produces a sufficient increase in centrifugal force to cause the spring-loaded governor plungers to protrude and close normally open overspeed switch S2. When overspeed switch S2 is closed, capacitor C3 discharges to fire squib "A" of Navol shutdown valve V5. When squib "A" is fired, the Navol shutdown valve closes to stop the flow of Navol to the decomposition chamber. The remainder of excessive speed shutdown is similar to normal shutdown.

3-146. COLD-RUN SHUTDOWN. A cold-run shutdown occurs when the combustion pot pressure does not increase to 400 psi within 10 seconds after the start of run or at any subsequent time during the run when the pot pressure drops below 400 psi. The combustion pot pressure will not reach 400 psi because the igniters fail, the fluid-delivery valves fail, or there is an interrupted flow of Navol and fuel within the combustion system.

3-147. In a cold-run shutdown, normally closed combustion pot pressure switch S3 remains closed or recloses because of insufficient combustion pot pressure. Therefore, when contacts 2 and 4 of delay-on-dropout relay K4 reclose 10 seconds after the start of a run, capacitor C4 discharges through contacts 2 and 4 of relay K4 and pot pressure switch S3 to fire squib "B" of the Navol shutdown valve V5. When squib "B" is fired, the Navol delivery valve closes and stops the flow of Navol to the decomposition chamber. The remainder of a cold-run shutdown is similar to normal shutdown.

3-148. ACR SHUTDOWN. A torpedo shutdown can also be initiated by the anti-circling-run (ACR) device. This shutdown occurs when the torpedo course heading changes more than 165° to left or right of the 0° gyro-angle setting. When this occurs, the ACR gyro pickoff switch closes the circuit made up of contacts 2 and 3 of delay-on-dropout relay K1 (ACR) and capacitor C3. At the instant the circuit is closed, capacitor C3 discharges to fire squib "A" of Navol shutdown valve V5. When squib "A" is fired, the Navol shutdown valve closes to stop the flow of Navol to the decomposition chamber. The remainder of ACR shutdown is similar to normal shutdown. After 55 seconds from start-of-run, contacts 2 and 3 of delay-on-dropout relay K1 (ACR) open to disable the ACR device and allow the desired left circle (enabling) to occur at the end of the straight portion of the torpedo run.

3-149. LOCATING SUNKEN TORPEDO. If the torpedo should sink at the end of a torpedo proofing or Fleet exercise run because of lack of positive buoyancy (or some other reason), it can be located by using a receiver that is capable of picking up the signal from Torpedo Locating Device (Pinger). The pinger is a self-contained, battery operated unit. It is manually started, then installed in an exercise head flange hole before loading the torpedo in the tube. Its operation produces a 37 kHz signal audible at a 1000 yard range using a receiver. The signal is

continuously generated until it is manually shut off or until its battery is exhausted. Refer to the manual furnished with each receiver for descriptive information. A second pinger is installed in the afterbody.

3-150. COMPLETE TORPEDO RUN. A normal run for a warshot torpedo is approximately 6.0 minutes from impulse. If an exercise run terminates before completing 4.9 minutes of normal propulsion, the torpedo will be negatively buoyant at shutdown. This means the torpedo will sink at the end of the run, regardless of whether or not the water ballast has been expelled from the exercise head. An exercise torpedo will, however, become positively buoyant by approximately 226 pounds when all expendables (water, alcohol, and Navol) and exercise head ballast have been either consumed or dumped.

3-151. The torpedo will shut down and begin pumping residual liquids (water and alcohol) overboard at any time the supply of oxygen or alcohol to the combustion system is interrupted; however, ballast will never be expelled from the exercise head until air-flask pressure has decayed to the point at which the air-release mechanism actuates. In the event that the propulsion system does not "light off" properly at the start of a run, the unit will run on the oxygen and alcohol flowing through the turbines, but at a depth approximately 10 feet shallower than set depth and at a greatly reduced speed for a period of 10 seconds. An abnormally started torpedo will shut down and sink 10 seconds after start. For all normal shots, end-of-run shutdown occurs when the Navol outlet port is no longer below the Navol level, thus allowing an air bubble to interrupt the Navol flow to the combustion system. This will occur when the tank has been drained dry or when the torpedo takes an extreme roll at a time when the liquid level in the Navol tank is low. The torpedo characteristically undergoes an extreme roll approximately 10 seconds after air-release mechanism actuation, when water ballast expulsion from the exercise head has been completed. At that time, a large quantity of air is exhausted from the exercise head and travels along the hull of the torpedo to the propellers, which then become partially unloaded, resulting in a violent roll.

3-152. The exercise torpedo starts its run with a negative buoyancy of approximately 970 lbs. During the run, the air-flask pressure is used up at rates which average 246 psi per minute, while the propulsion system consumes expendables at a rate of 80 pounds per minute. When air-release mechanism actuation occurs, approximately 577 pounds of water ballast is expelled from the exercise head.

3-153. To achieve positive buoyancy by the time the torpedo shuts down, the air-release mechanism must be set to between 1425 and 1475 psi and the air flask must be charged to between 2750 and 2850 psi for the start of the torpedo run.

3-154. An optimum exercise run of 5.7 minutes by a torpedo having an initial flask charge of 2800 psi and air-release mechanism setting of 1450 psi will have used the required 1350 psi differential to initiate the air-release mechanism by the time the torpedo has run 5.5 minutes. Immediately following the 10-second head blow duration, the torpedo will shut down as a result of the roll caused by air discharging from the exercise head. At shutdown, the torpedo will be positvely buoyant by approximately 63 pounds. Venting of tanks and dumping of an additional 184 pounds of air, water, fuel, and possibly a small quantity of Navol, will continue for approximately 9 minutes after shutdown until the full 266 pounds of positive buoyancy is attained.

3-155. If the air flask pressure does not decay at a rate of 246 psi per minute during the run, the air-release mechanism will not actuate before the torpedo has run for 6.0 minutes. In this case the Navol tank will run dry and the torpedo will shut down and sink. The torpedo will later come to the surface (provided a depth of 300 feet is not exceeded) when the air-release mechanism does actuate and expel water ballast from the exercise head. The air-release mechanism setting and air-flask pressure charge must be conscientiously maintained within specified limits or various end-of-run sinkers will result.

INDEX

DISTRIBUTION LIST

SNDL Part 1 (No. 105, 1 March 1973) and Part 2 (No. 55, 15 February 1973), two copies each unless otherwise indicated.

21A (CINCPACFLT and CINCLANTFLT only); 22 (1 copy); 24F; 24G (less REP); 26F (COMOPTEVFOR only); 28K1 (less SUBRON 14, 1 copy each, except COMSUBRON-4, 3 copies); 29S1 (SS-426, 3 copies, 566 1 copy, others 2 copies each); 29S4 (SSN-575, 579, 583, 606, 637, 653, 662, 663, 679, 3 copies each, 684 5 copies, others 2 copies each); 29S8 (SSBN-601 and 625, 3 copies, others 2 copies each); 31F; 32DD; 32EE (less ASR-9); A3 (OP-201, 1 copy); C4F9 (AUTEC, Attn: Library, 1 copy); FA6; FA7 (Key West and Roosevelt Roads, WSAT); FA10 (8 copies); FB13 (4 copies); FB30 (Yokosuka only); FB31 (Subic Bay only, 3 copies); FC4 (Sigonella, 1 copy); FC7; FF5; FKA1D (ORD-0832, ORD-541, ORD-0632, 1 copy each); FKA6A8; FKA6A9 (Attn: Library); FKA6A15 (8 copies); FKL1 (Mare Island, Code M245F, 3 copies, Philadelphia Code 249B, 5 copies, Pearl Harbor Code 246-P, 5 copies, Portsmouth, Code 190, 1 copy); FKL2 (Groton Code 249, 4 copies and Pascagoula, 3 copies); FKL8 (Code 6161E); FKM13 (Code 88232, 1 copy); FKP1A (Earle Code 0112 and Oahu only); FKP1B (Yorktown Code 20323, 7 copies, Seal Beach Code 0114, 8 copies, Charleston 3 copies, Concord, 2 copies); FKP1C (Pacific only, 1 copy); FKP1E (11 copies); FKP1J (Louisville, Code TDD, 3 copies); FKP3A (Pomona, Attn: Metrology Library, 1 copy); FKP5A; FKR7A (Pacific only); FR4 (1 copy); FT5; FT23 (4 copies); FT24 (Newport 16 copies, Norfolk 1 copy); FT30 (Orlando, 55 copies, Great Lakes, BE/E School, 1 copy); FT38 (16 copies); FT45 (Attn: Library Div., 4 copies); FT46 (1 copy); FT47; FT54 (21 copies); FT54 (Weapons Div., Sub School Staff, Bldg. 83, 5 copies); FT61 (1 copy); FT64 (1 copy); APL/UW, Seattle; Automation Industries, Vitro Laboratories, 14000 Georgia Ave., Silver Spring, MD 20910.

Requests for additional copies of NAVORD OP 3358 (FIRST REVISION) Volume 1 should be addressed to the Commanding Officer, Naval Publications and Forms Center, 5801 Tabor Avenue, Philadelphia, PA 19120.

Want a Better Manual?

... for fast results without red tape
try the Publication Review postcard!

Here's how it works—

Fill in one of the attached postcards, stating in your own words any ideas, suggestions, or recommendations for improving this manual. Drop the card in the mail; no postage is necessary. You will receive an acknowledgement as soon as your comments have been reviewed by the people concerned.

All adopted suggestions will be incorporated in the next revision of the manual. Recommended technical changes or errors brought to our attention will be taken care of immediately by means of change sheets.

What we want—

Technical errors in text or illustrations—additions to the text—comments on arrangement of material — omissions of vital procedures—clarity of writing — adequacy of trouble-shooting charts, check lists, tables, etc.

IMPORTANT — Be sure to fill in your name, ship or station, and address in the space provided on this side of the postcard. Additional copies of this form may be obtained from addressee.

PUBLICATION REVIEW

PUBLICATION NO.	REVISION NO.	TITLE

SIRS: I have read the publication and find that it is:

☐ In Error   ☐ Incomplete   ☐ Difficult to Understand   ☐ Poorly Arranged

Specific Comments: [For lengthy reports, use NAVWEPS Form 8510/3 (RUDTORPE).]

SUGGESTOR (Name and Rank)

DATE

ADDRESS (Ship or Station)

PUBLICATION REVIEW

PUBLICATION NO.	REVISION NO.	TITLE

SIRS: I have read the publication and find that it is:

☐ In Error   ☐ Incomplete   ☐ Difficult to Understand   ☐ Poorly Arranged

Specific Comments: [For lengthy reports, use NAVWEPS Form 8510/3 (RUDTORPE).]

SUGGESTOR (Name and Rank)

DATE

ADDRESS (Ship or Station)

Our best guide is your comments

... you can help your Navy improve its manuals by giving us your ideas

No matter how unimportant your suggestions or recommendations for improvement of this manual may seem to you, send them along. The Navy welcomes every contribution and will consider each for possible use in the next revision of this manual.

By filling in the attached card and mailing it to NUSC, the reader is providing the Technical Manual branches of the Navy with invaluable information for improving the adequacy and reliability of future torpedo publications.

NAVAL UNDERWATER SYSTEMS CENTER
NEWPORT LABORATORY
NEWPORT, R. I. 02840

POSTAGE AND FEES PAID
NAVY DEPARTMENT

OFFICIAL BUSINESS

COMMANDING OFFICER
NAVAL UNDERWATER SYSTEMS CENTER
NEWPORT LABORATORY
NEWPORT, RHODE ISLAND 02840

TECHNICAL INFORMATION

NAVAL UNDERWATER SYSTEMS CENTER
NEWPORT LABORATORY
NEWPORT, R. I. 02840

POSTAGE AND FEES PAID
NAVY DEPARTMENT

OFFICIAL BUSINESS

COMMANDING OFFICER
NAVAL UNDERWATER SYSTEMS CENTER
NEWPORT LABORATORY
NEWPORT, RHODE ISLAND 02840

TECHNICAL INFORMATION

Before you mail this postcard...

←

MAKE SURE YOUR COMMENTS ARE UNCLASSIFIED

●

If the information on this card is classified, mail it in accordance with OPNAV Instruction 5510.1B, Department of the Navy Security Manual for Classified Information.

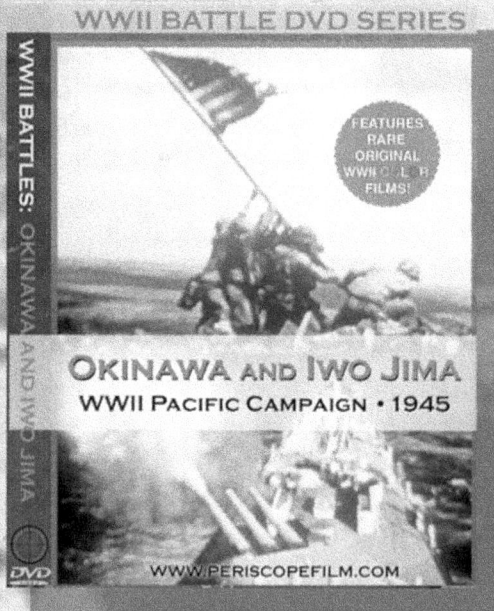

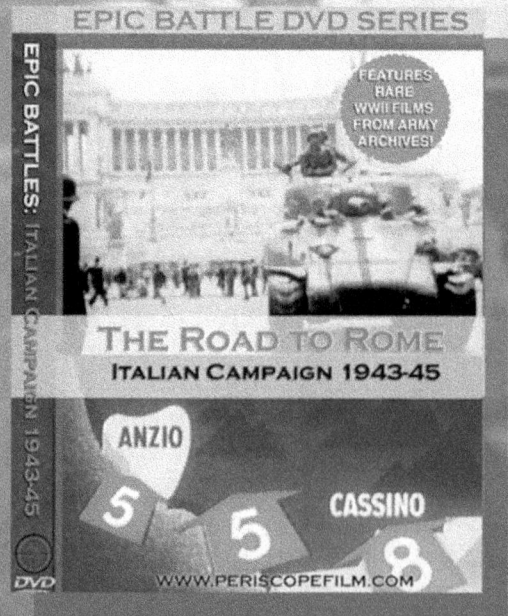

www.ingramcontent.com/pod-product-compliance
Lightning Source LLC
Chambersburg PA
CBHW080514110426
42742CB00017B/3109